Techniques and Applications of Biomedical AI

Mohamed Hammad
Menoufia University, Egypt & Prince Sultan University, Saudi Arabia

Hatem Abdelkader
Menoufia University, Egypt

IGI Global
Scientific Publishing
Publishing Tomorrow's Research Today

Vice President of Editorial Melissa Wagner
Director of Acquisitions Mikaela Felty
Director of Book Development Jocelynn Hessler
Production Manager Mike Brehm
Cover Design Jose Rosado

Published in the United States of America by
IGI Global Scientific Publishing
701 East Chocolate Avenue
Hershey, PA, 17033, USA
Tel: 717-533-8845
Fax: 717-533-7115
Website: https://www.igi-global.com E-mail: cust@igi-global.com

Library of Congress Cataloging-in-Publication Data

Names: Hammad, Mohamed, 1990- editor | Abdelkader, Hatem, 1968- editor
Title: Techniques and applications of biomedical AI / edited by Mohamed Hammad, Hatem Abdelkader.
Other titles: Techniques and applications of biomedical artificial intelligence
Description: Hershey, PA : IGI Global Scientific Publishing, [2025] | Includes bibliographical references and index. | Summary: "The primary objective of this book is to bridge the gap between AI capabilities and practical applications in biomedical research by focusing on prompt engineering. The book will explore not only the technical and methodological aspects of designing prompts but also the ethical, regulatory, and educational dimensions that accompany their use"-- Provided by publisher.
Identifiers: LCCN 2025002906 (print) | LCCN 2025002907 (ebook) | ISBN 9798337305486 hardcover | ISBN 9798337305493 paperback | ISBN 9798337305509 ebook
Subjects: LCSH: Artificial intelligence--Medical applications | Artificial intelligence--Biological applications
Classification: LCC R859.7.A78 .T43 2025 (print) | LCC R859.7.A78 (ebook) | DDC 610.285--dc23/eng/20250613
LC record available at https://lccn.loc.gov/2025002906
LC ebook record available at https://lccn.loc.gov/2025002907

British Cataloguing in Publication Data
A Cataloguing in Publication record for this book is available from the British Library.

Table of Contents

Detailed Table of Contents

This chapter aims to develop an AI-powered chatbot designed to provide personalized Ayurvedic and home remedy solutions for individuals suffering from sinusitis. Sinusitis, an inflammation of the sinuses, is a common condition that can cause severe discomfort, including headaches, nasal congestion, and facial pain. This chatbot leverages artificial intelligence to analyze user inputs and recommend tailored Ayurvedic remedies, lifestyle changes, and home treatments based on the individual's specific symptoms and health history. The system's continuous learning capability enables it to refine and enhance its advice based on medical data, ensuring its advice is relevant and up-to-date. By providing instant, accessible, and reliable guidance, the chatbot aims to empower users to manage their sinusitis effectively with natural treatments. The project demonstrates the potential of AI in bridging the gap between traditional medicine and modern healthcare, offering a user-friendly interface that supports wellness and informed decision-making.

Ultrasound tomography is a widely adopted diagnostic imaging modality that has undergone rapid advancements in detecting sub-wavelength tumors without the need for ionizing radiation, making it a preferable alternative to the current gold standard of X-ray mammography. This imaging technique relies on an inverse scattering algorithm that utilizes material properties, including sound contrast or attenuation, to detect and identify small targets. In this work, we propose the use of the distorted Born iterative (DBI) method, which is an efficient diffraction tomography approach based on the first-order Born approximation, to improve the biomedical signal processing techniques in ultrasound tomography. Specifically, we investigate the efficacy of various signal processing techniques, including interpolation, dual-frequency, multi-resolution, and multi-frequency DBI (MR-MF-DBI), through numerical simulations. This study serves as a proof-of-concept and will be further expanded using experimental data before clinical application.

Cardiovascular disease (CVD) remains a significant global health concern, influenced by factors like lifestyle, lipid profiles, and genetics. Early prediction is vital for effective prevention and management. This study develops predictive models for CVD risk by combining demographic, health, genetic, and imaging data, utilizing machine learning (ML) and deep learning (DL) techniques. The main objective is to enhance risk prediction accuracy by analyzing X-ray image data through convolutional neural networks (CNNs). Additionally, the study explores the integration of structured patient data (demographics, lipid profiles, genetics) with imaging features to provide a comprehensive CVD risk assessment. The dataset, consisting of 264 patients, includes diverse factors like lifestyle and health indicators. Data preprocessing involved normalization, resizing, and feature selection through techniques like LASSO. ML models (SVM, RF, DT) achieved up to 95% accuracy, while CNNs analyzing X-ray images reached 100% accuracy.

This chapter provides a comprehensive exploration of machine learning applications in medical image analysis across various imaging modalities including X-ray, computed tomography (CT), magnetic resonance imaging (MRI), ultrasound, and positron emission tomography (PET). We begin by examining the fundamental principles of medical imaging and traditional image processing techniques before delving into machine learning paradigms—supervised, unsupervised, and semi-supervised learning approaches—that have been successfully applied to healthcare imaging challenges. The evolution from conventional machine learning algorithms to advanced deep learning architectures is thoroughly discussed, with particular emphasis on convolutional neural networks (CNNs) and their specialized variants such as U-Net for medical image segmentation. The chapter presents detailed applications across clinical domains, including disease detection, organ segmentation, computer-aided diagnosis, and treatment response prediction.

This chapter focuses on identifying the potential of prompt engineering to revolutionise health care with the help of applied research and case studies. With the help of new generative models, such as GPT-4, healthcare specialists work on clinical problems with the help of better diagnostics, improved interaction with patients, and more effective treatment plans. This chapter describes the strategies for calibrating prompts to produce context-aware, reliable responses that can support real-time decision-making in clinical practice. Prompt engineering enhances the health of clinicians since the software manages routine jobs. This chapter also looks at its usage in medical education using AI-based case scenarios, which help grasp intricate clinical ideas and reasoning abilities. Thus, prompt engineering can help healthcare professionals increase diagnostic performance, engage patients more effectively, and optimise the work process. This chapter establishes the need for prompt engineering as one of the most essential tools in the current and future healthcare field.

The integration of fuzzy logic and related approaches in healthcare has emerged as a promising avenue for addressing the inherent complexities and uncertainties in medical data and decision-making processes. This review chapter explores the application of fuzzy logic in various domains of healthcare, including medical diagnosis, treatment planning, patient monitoring, and healthcare management. By accommodating imprecise information, incomplete data, and subjective interpretations, fuzzy approaches offer a flexible framework for modelling and analyzing healthcare systems. The chapter examines the principles of fuzzy logic, its implementation in healthcare contexts, and the potential benefits it brings, such as enhanced decision support, personalized treatment strategies, and improved patient outcomes. Furthermore, challenges and limitations associated with the adoption of fuzzy approaches in healthcare are discussed, along with future research directions. This chapter serves as a comprehensive overview of the role of fuzzy logic in shaping the future of healthcare.

Preface

INTRODUCTION

The integration of artificial intelligence (AI) into biomedical research and healthcare systems marks a transformative era in medical science. The rapid advancement of machine learning, signal processing, and generative models has opened new pathways for disease detection, treatment optimization, and personalized care. This edited volume, *Techniques and Applications of Biomedical AI*, brings together diverse innovations from around the world that explore how AI technologies are being applied to address some of the most critical challenges in medicine today. From diagnostic imaging to holistic wellness solutions, these contributions reflect the interdisciplinary nature of the field and the collaborative spirit of AI-driven healthcare advancement.

The aim of this book is not only to present cutting-edge research but also to provide readers with an accessible foundation in the methodologies and implications of AI in biomedicine. Whether it is the early detection of skin cancer using 3D imaging, the use of chatbots to bridge traditional medicine and digital health, or the optimization of diagnostic performance through prompt engineering, each chapter is designed to deepen understanding and spark further inquiry. The work showcased here serves academic, clinical, and technological audiences alike, offering insights that are both practical and forward-looking.

With contributions from researchers and practitioners across a variety of institutions and countries, this book highlights the global nature of innovation in biomedical AI. As AI continues to evolve, the insights provided in this volume will remain relevant for guiding future studies, informing clinical decisions, and influencing health policies. Our hope is that readers leave with both knowledge and inspiration to contribute to this exciting and rapidly developing field.

CHAPTER OVERVIEW

Chapter 1: AI Skin Cancer Detection with 3D-TBP

This chapter presents a deep learning-based system that analyzes 3D Total Body Photography images to detect skin cancer types such as melanoma, basal cell carcinoma, and squamous cell carcinoma. By employing ResNet architecture and stratified group k-fold cross-validation, the model achieves a robust classification of malignant versus benign lesions. It underscores the value of early diagnosis through AI-powered imaging tools.

Chapter 2: AI-Powered Chatbots for Managing Sinusitis – Integration of Ayurvedic and Home Remedies

This chapter introduces an AI-driven chatbot designed to provide natural, personalized treatment recommendations for sinusitis sufferers. Drawing from Ayurvedic principles and home remedies, the chatbot integrates user-specific health data to offer targeted advice. It demonstrates how AI can support complementary and integrative healthcare, promoting accessibility and user empowerment.

Chapter 3: Biomedical Signal Processing Techniques for Enhanced Distorted Born Iterative Method in Ultrasound Tomography

Focusing on ultrasound tomography, this chapter enhances tumor detection capabilities using the distorted Born iterative method. Through simulation-based evaluation of interpolation and multi-frequency signal processing techniques, the study contributes to the development of more precise, radiation-free imaging methodologies. This research is a critical step toward safer, more accurate diagnostics.

Chapter 4: Examining Personal Lifestyle Factors and Lipid Profile on Cardiovascular Disease Risk Prediction

This chapter explores the predictive modeling of cardiovascular disease using structured and unstructured data. By merging lifestyle information, lipid profiles, and imaging data into a machine learning framework, the chapter highlights how CNNs and traditional models can be used to reach near-perfect predictive accuracy. It emphasizes the potential of AI in proactive, personalized healthcare.

Chapter 5: Machine Learning for Image Processing in Healthcare

Covering multiple medical imaging modalities, this chapter provides a comprehensive survey of how machine learning techniques—from traditional algorithms to modern deep learning networks—are improving diagnosis and treatment planning. Particular focus is given to CNN-based segmentation, classification, and disease detection, outlining how AI supports clinical decision-making across various conditions.

Chapter 6: Optimising Healthcare Outcomes with Prompt Engineering: A Transformative Approach

Prompt engineering, especially within generative AI models like GPT-4, is the focus of this chapter. It delves into methods for structuring prompts to improve diagnostics, enhance patient engagement, and streamline medical education. This research affirms the utility of natural language processing in reducing clinical workload while maintaining accuracy and empathy in healthcare delivery.

Chapter 7: The Fuzzy Frontier – Exploring the Potential of Uncertainty in Healthcare

The final chapter addresses the use of fuzzy logic to manage the uncertainties and ambiguities inherent in medical data. It explores applications in diagnosis, patient monitoring, and health management, showing how fuzzy models enhance decision-making under imprecise conditions. The chapter also discusses the limitations and future opportunities for integrating fuzzy logic into clinical systems.

CONCLUSION

As healthcare continues to grapple with increasing complexity, AI offers solutions that are not only technologically advanced but also deeply human-centric. The contributions in this volume illustrate how AI, when applied thoughtfully and rigorously, can enhance diagnostics, personalize treatment, and improve patient engagement. They show that the future of medicine lies not only in machines but in the intelligent synergy between human insight and computational power.

This book reflects a paradigm shift—one where interdisciplinary collaboration, open innovation, and ethical application of technology converge. While each chapter delves into a specific use case, the broader narrative underscores the importance of trust, interpretability, and inclusivity in AI-driven healthcare systems. These themes are essential as we build a healthcare landscape that is both effective and equitable.

We extend our gratitude to all contributors who have shared their expertise and to the global scientific community continually pushing the boundaries of what AI can achieve. *Techniques and Applications of Biomedical AI* serves as a testament to the possibilities that lie ahead, and we hope it encourages further research, collaboration, and responsible innovation in the biomedical field.

Acknowledgment

We sincerely thank all those whose contributions made this edited book possible.

Our heartfelt gratitude goes to the authors for their insightful chapters, which have enriched this volume with depth and value. We also thank the reviewers for their thoughtful feedback, which significantly improved the quality and rigor of the work. Special thanks to the editorial and publishing teams for their professionalism and dedication throughout the process. Your commitment ensured the successful realization of this project. We are also grateful to our families, friends, and colleagues for their unwavering support during the many hours devoted to this book. Your encouragement has been a constant source of strength.

To our readers—thank you for your interest. We hope the ideas within these pages inspire meaningful dialogue and future innovation.

Mohamed Hammad and Hatem Abdelkader

A Personal Note from Dr. Mohamed Hammad:

I would like to express my deepest gratitude to my sons, Hamza and Yousef, whose understanding and quiet strength have been a constant source of motivation throughout this endeavor. Their patience during my extended hours of work has been invaluable, and their presence a grounding force.

To my wife, Asmaa Maher—your unwavering support, patience, and dedication have made this journey possible. Your willingness to manage countless responsibilities with grace allowed me the focus needed to complete this work. I am sincerely thankful for your steadfast encouragement and belief in my efforts.

With deep gratitude,

Mohamed Hammad

Chapter 1
AI Skin Cancer Detection With 3D-TBP

V. Santhi
https://orcid.org/0000-0002-4274-7474
Vellore Institute of Technology, Chennai, India

Savithri Nair
Vellore Institute of Technology, Chennai, India

Gazal Agrawal
https://orcid.org/0009-0009-9400-5905
Vellore Institute of Technology, Chennai, India

Sakshi Dodiya
Vellore Institute of Technology, Chennai, India

ABSTRACT

Early detection of skin cancer, including melanoma, basal cell carcinoma, and squamous cell carcinoma, is crucial for improving patient outcomes. This chapter presents an AI-based approach to address these challenges using a novel dataset of 3D Total Body Photography (3D-TBP) images. Our goal is to differentiate histologically confirmed malignant skin lesions from benign ones, thereby extending the benefits of automated skin cancer detection to broader populations. To reduce overfitting and ensure better generalization, we employed a stratified group k-fold cross-validation strategy, maintaining balanced representation of malignant and benign cases while preserving patient-level independence across folds. We utilized a ResNet-based deep learning architecture to extract and learn complex patterns from image data and the model was evaluated using the partial area under the Receiver Operating Characteristic curve (pAUC) with 80% true positive rate. Our

DOI: 10.4018/979-8-3373-0548-6.ch001

approach demonstrated competitive performance, indicating its potential in early skin cancer detection.

1. INTRODUCTION

Skin cancer is among the most prevalent cancers worldwide, impacting millions annually. It occurs when skin cells grow abnormally, often due to extended exposure to ultraviolet (UV) rays from sunlight or artificial sources such as tanning beds. Additionally, genetic predispositions and environmental factors can influence its development. The primary types of skin cancer include basal cell carcinoma (BCC), squamous cell carcinoma (SCC), and melanoma, each with distinct features and varying levels of risk.

Basal cell carcinoma is the most common type and is primarily caused by UV radiation. It often appears as raised, shiny nodules or red patches on sun-exposed areas, like the face and neck. Although BCC typically grows slowly and seldom spreads, it can lead to considerable damage if not treated. Squamous cell carcinoma, the second most common type, also develops from UV exposure and typically presents as rough, scaly red patches or open sores that may bleed or crust over. While SCC can be more aggressive and has a higher potential to spread, early detection usually results in a positive prognosis. Melanoma, though less common, is the most dangerous form of skin cancer. It starts in pigment-producing cells and is known for its ability to grow quickly and metastasize. Melanoma often appears as an irregular mole with asymmetrical borders, varying colors, or changes in size. While it can develop anywhere on the body, it is most common in sun-exposed areas such as the back, legs, and arms. Early detection of melanoma can significantly improve survival rates, but if left untreated, it can spread rapidly to other organs.

The incidence of melanoma has been increasing globally, with approximately 325,000 new cases diagnosed in 2023, leading to over 57,000 deaths. This increase is linked to higher UV exposure, lifestyle changes, and improved screening methods. Early detection of skin cancer, especially melanoma, is crucial for effective treatment and improved outcomes. When caught in its early stages, melanoma has a high survival rate, but if diagnosed late, it becomes harder to treat and more likely to spread. Regular skin checks, self-examinations, and the use of technologies like dermoscopy and 3D Total Body Photography can help detect skin cancer earlier. AI-driven tools are also showing promise in enhancing early detection and improving diagnostic accuracy, making them invaluable in the fight against skin cancer. Raising awareness about sun protection, recognizing symptoms, and encouraging routine skin checks are essential for reducing the impact of skin cancer worldwide.

2. LITERATURE REVIEW

A study conducted at the University Hospital Carl Gustav Carus aimed to explore the effectiveness of 3D Total Body Photography (3D-TBP) in identifying non-melanoma skin cancer (NMSC) among 129 patients treated from April 21, 2021, to June 16, 2022. This research sought to evaluate how well 3D-TBP could detect and diagnose NMSC and its precursors, comparing its performance to dermoscopy and identifying its strengths and weaknesses. 3D-TBP creates a comprehensive 3D skin model by capturing a series of images with a specialized camera system, which allows clinicians to examine areas of interest from multiple angles. Unlike dermoscopy, a traditional method that uses a handheld device to magnify and illuminate skin lesions with high-resolution detail, 3D-TBP offers a broader, three-dimensional view of the entire body surface. The findings revealed that 3D-TBP demonstrated strong sensitivity and specificity in diagnosing NMSC. Its ability to provide multi-angle views of lesions helped clinicians gain valuable supplementary insights, such as observing changes in lesion shape and size over time. This made it especially useful for structured assessments of skin changes. However, the study noted that 3D-TBP's lower image resolution compared to dermoscopy resulted in slightly reduced diagnostic accuracy. Additionally, biopsies were only performed for in-situ squamous cell carcinoma (SCC) and superficial basal cell carcinoma (BCC) when there was diagnostic uncertainty, which may have affected the sensitivity and specificity results. Overall, the study highlighted both the strengths and limitations of 3D-TBP. Its capability to enable structured, multi-angle evaluations of skin lesions holds promise for early and accurate diagnosis, while its lower image resolution remains a drawback. Further research is necessary to better understand how 3D-TBP compares to dermoscopy and to address its limitations. If refined, 3D-TBP could significantly improve early detection and management of NMSC, benefiting patient care through timely and precise interventions (Hobelsberger et al., 2023).

Recent advancements in imaging technologies have demonstrated substantial potential for enhancing the diagnosis of skin cancer, focusing on non-melanoma and melanoma types. A notable prospective study conducted at the University Hospital Carl Gustav Carus evaluated the diagnostic utility of 3D Total Body Photography (3D-TBP) in detecting non-melanoma skin cancer (NMSC) across 129 patients between April 21, 2021, and June 16, 2022. This research compared 3D-TBP's performance with dermoscopy, revealing that while 3D-TBP exhibited high sensitivity and specificity, it was slightly less accurate than dermoscopy due to its lower image resolution. Nevertheless, 3D-TBP's capacity to provide multi-angle visualizations of lesions proved beneficial for structured diagnostics and monitoring lesion changes over time. The study underscores the need for further exploration to validate 3D-TBP's diagnostic capabilities and address its inherent limitations. Parallel advancements in augmented intelligence for dermatological applications have demonstrated the efficacy of combining two-dimensional (2D) and three-dimensional (3D) total body photography convolutional neural networks (TBP CNN) for classifying and analyzing melanocytic lesions within high-risk melanoma populations. Utilizing CNN models, 3D TBP CNN exhibited superior performance relative to 2D TBP CNN in lesion detection, achieving a higher correlation coefficient across repeated measurements and demonstrating reduced variability in scoring. Despite this, both models exhibited notable limitations, including overcounting lesions and inconsistencies in TBP capture due to technical factors, limiting their clinical practicality. These challenges underscore the necessity for model refinement to enhance accuracy and dependability in everyday clinical settings. Nevertheless, the integration of AI-driven TBP approaches in dermatology offers significant promise for advancing early skin lesion detection and optimizing screening outcomes. This dual emphasis on refined imaging and AI technologies reflects a pivotal shift towards more accurate, comprehensive skin cancer diagnostics in high-risk populations (Cerminara et al., 2023).

Recent research has been exploring how Total Body Photography (TBP) combined with Artificial Intelligence (AI) could transform skin cancer diagnosis by offering a more holistic approach. One such study involved 500 patients from Australia and Spain, where data collected from TBP imaging, sun behavior surveys, medical history, and genetic risk factors were combined to build a comprehensive dataset for AI analysis. The aim was to develop an AI system that doesn't just focus on individual skin lesions but considers the whole patient, factoring in their medical history and genetic background to improve the accuracy of skin cancer detection. The dataset was created through temporal full-body imaging and DNA samples, with careful anonymization to protect patient privacy. The strength of this approach lies in its ability to evaluate skin cancer risk on a more personal level, considering not only the lesions but also the patient's overall health context. TBP provides a full-body view, helping clinicians assess potential risks from multiple angles, while AI can analyze the data more efficiently and accurately than traditional methods. However, the study doesn't provide in-depth information about the AI algorithms used or how effective the approach would be in real-world clinical settings. Additionally, challenges like the cost of TBP equipment and the need for standardized image annotations have not been fully addressed. While this combination of TBP and AI holds great promise for revolutionizing skin cancer screening, more research is needed to fine-tune the AI algorithms and solve the practical issues that might arise in everyday clinical use (Primiero et al., 2024).

In recent years, deep learning models, particularly convolutional neural networks (CNNs), have made significant strides in skin cancer detection. A study focused on using the ResNet-152 CNN model, which is considered a cutting-edge deep learning model, aimed to enhance skin cancer classification. The researchers fine-tuned this model with the pretrained ImageNet network to analyze skin lesions effectively. To prepare the images for training, lesions were cropped and resized to a standard 224×224 pixels. The researchers carefully optimized the model's sensitivity and specificity, determining the best threshold that maximized the Youden index, which helps assess the accuracy of diagnostic tests. A key aspect of their evaluation process was drawing a receiver operating characteristic (ROC) curve to compare the model's outputs against the chosen threshold, assessing its effectiveness in detecting skin cancer. To ensure the robustness and generalizability of the model, the study incorporated several large datasets, including the Asan dataset (17,125 images), the Atlas dataset (3,820 images), and the Edinburgh dataset (1,300 images). These datasets provided a wide variety of benign and malignant lesions, allowing for a detailed comparison of the deep learning model's performance with a panel of dermatologists. The researchers randomly selected 25 images of each malignancy and 20 images of each benign lesion from the Asan and Edinburgh datasets for this comparison. This comprehensive approach demonstrated the potential of deep learning in skin cancer detection, particularly in large-scale clinical datasets. However, while the study presented impressive methodology, it lacked key performance metrics like accuracy, precision, recall, or F1 score, which are essential for understanding the model's real-world efficacy. Additionally, details on the computational resources used for training the model were not provided, leaving some questions about the operational feasibility of the approach. The study also did not discuss how it addressed class imbalance, a common challenge in medical image classification. Despite these gaps, the research underscores the importance of using clinical datasets and modern technological advancements to push the boundaries of AI in dermatology. Going forward, more research is needed to address these limitations and further refine AI models for real-world use in skin cancer detection (Han et al., 2018).

Skin cancer, particularly melanoma, is a critical health concern due to its aggressive nature and high mortality if not diagnosed early. Traditional standard-of-care (SOC) examinations primarily rely on visual inspection and dermoscopy, but these can miss subtle or complex melanomas, especially in high-risk patients. To improve diagnostic accuracy, a three-step advanced imaging program was tested on 410 melanoma patient images, combining 3D total body photography (3D-TBP), digital dermoscopy (DD), and reflectance confocal microscopy (RCM). Each modality offers unique strengths: 3D-TBP creates a comprehensive visual map of the entire body, allowing for effective lesion tracking and detection of new growths over time. DD provides high-resolution images of lesions, revealing fine details in pigmented and vascular structures that aid in differentiating benign from malignant growths. RCM delivers real-time, non-invasive cellular-level imaging, capturing intricate sub-surface structures and enhancing diagnostic precision. Individually, RCM demonstrated superior ability in identifying cellular-level abnormalities, making it an invaluable tool for distinguishing melanomas. However, integrating all three modalities proved most effective, as each identified unique lesion characteristics that could be missed by the others, resulting in enhanced diagnostic sensitivity and specificity. Despite the promise of this combined approach, challenges persist. The high cost of equipment, maintenance, and specialized training needed to operate and interpret these technologies limits their accessibility, particularly in resource-constrained settings. The combination of these techniques can also extend examination time, potentially impacting clinical workflow and patient throughput. Future efforts should focus on making these advanced imaging technologies more cost-effective and user-friendly, thereby increasing accessibility in various clinical settings. Integrating artificial intelligence (AI) into image analysis could streamline lesion assessments, improve diagnostic consistency, and reduce interpretation variability. Research into optimizing clinical workflows and enhancing training programs is essential for facilitating routine integration of these modalities. Additionally, large-scale validation studies are necessary to confirm the long-term impact of this multimodal approach on patient outcomes and healthcare efficiency. Overall, combining 3D-TBP, DD, and RCM holds significant potential for improving melanoma detection and patient care by offering a comprehensive, synergistic diagnostic strategy that surpasses the capabilities of SOC examinations alone (Gellrich et al., 2024).

The growing prevalence of melanoma highlights the urgent need for effective and accessible early detection methods. Among emerging solutions, CE-certified smartphone applications have garnered attention as tools for initial skin lesion assessments, offering users a convenient way to gauge melanoma risk. These apps utilize artificial intelligence (AI) and advanced image processing techniques, employing 2D and 3D convolutional neural networks (CNNs) to analyze lesion images and classify patients into three risk categories: low, moderate, and high. Those identified as moderate or high-risk are advised to seek further evaluation from a dermatologist. The appeal of these apps lies in their accessibility and user-friendly design, enabling individuals to perform preliminary assessments from the comfort of their homes. By leveraging large datasets, the AI-driven algorithms are capable of detecting visual patterns and characteristics indicative of melanoma with considerable accuracy. This technology holds promise, especially in regions where access to dermatologists is limited, potentially empowering more people to detect suspicious lesions earlier. However, challenges remain. A notable concern is over-detection—the tendency for these apps to flag benign lesions as potentially malignant, which can lead to unnecessary worry for users and place a strain on healthcare systems due to increased referrals. Unlike dermatologists, who bring clinical expertise and context-based judgment to their evaluations, AI apps rely solely on image analysis, which can result in misclassifications or false positives. While these tools provide valuable initial insights, they cannot yet match the depth of human expertise. Despite these limitations, the advantages are clear. The app's ability to offer a non-invasive, readily available risk assessment can enhance public awareness and prompt earlier consultations with healthcare professionals. This could play a crucial role in improving early melanoma detection and outcomes. However, there are potential downsides—misdiagnosis risks, user misunderstandings of results, and reduced sensitivity for complex or atypical cases that may need an expert's eye. Moving forward, efforts should focus on refining the algorithms to reduce over-detection rates, integrating human oversight into AI-driven processes to ensure accuracy, and educating the public on the app's intended use as a preliminary tool rather than a definitive diagnosis. Expanding clinical trials and comparative studies against dermatologist-led assessments will help validate and shape these technologies for safe, effective use in healthcare settings. In summary, while CE-certified smartphone apps for melanoma screening hold significant promise as supplementary tools, their optimal value lies in careful deployment alongside clinical guidance, ensuring they empower users without compromising on accuracy or safety (Jahn et al., 2022).

The growing use of artificial intelligence (AI) in healthcare, especially for melanoma screening, has sparked significant interest due to its potential to improve diagnostic accuracy and efficiency. A recent prospective study compared how doctors and AI systems perform in detecting skin cancer, specifically melanoma, and examined the perspectives of both patients and dermatologists on AI's role in diagnosis. The study found that AI, particularly tools using convolutional neural networks (CNNs), could analyze skin images with remarkable precision. These AI systems are trained on large datasets of melanoma images, allowing them to detect patterns that might be overlooked by the human eye. This has led to a growing belief that AI can be a helpful tool for identifying suspicious lesions, particularly in areas with limited access to dermatologists. For patients, AI offers a level of convenience and accessibility that traditional visits to the doctor may not. Many people find AI-driven tools appealing because they can quickly assess their skin conditions without needing to see a specialist right away. However, while patients may rely more on AI for its efficiency, dermatologists still trust their clinical judgment and experience when diagnosing melanoma. Despite AI's growing capabilities, doctors continue to emphasize their expertise, recognizing that melanoma detection requires a nuanced understanding of skin conditions that AI cannot fully replicate.

The study concluded that while patients may be inclined to trust AI for its convenience, dermatologists remain focused on using their professional judgment alongside AI tools. This highlights the idea that AI should complement—not replace—human expertise in healthcare. In the future, combining AI's efficiency with the in-depth knowledge of healthcare professionals could lead to more effective melanoma screening and earlier diagnosis. Additionally, educating patients on the strengths and limitations of AI will be essential for ensuring it is used appropriately in healthcare settings (Goessinger et al., 2024).

The use of AI in melanoma detection has been rapidly advancing, with the Australian experiment offering an innovative approach to skin cancer screening through a 3D total-body photography network. This system combines detailed skin images with questionnaires about a patient's family history of skin cancer, lifestyle habits, and other factors, creating a holistic profile that helps improve diagnosis.

The process begins with capturing high-resolution 3D images of the patient's skin. Along with these images, the patient fills out a questionnaire that provides important details about their personal and family history of skin cancer, UV exposure, and other relevant health information. AI models then analyze both the images and the questionnaire data, producing a textual report that highlights any areas of concern, such as suspicious lesions that might indicate melanoma. After the AI generates this report, dermatologists review it to decide whether any of the flagged areas are cancerous or require further investigation. The AI serves as a supportive tool, helping doctors identify the most concerning lesions and improving the efficiency of the screening process. This system allows dermatologists to focus on the most critical cases while also benefiting from the AI's ability to quickly analyze large amounts of data. This approach has clear advantages, especially in terms of speeding up the screening process and assisting doctors with decision-making. However, challenges still exist, such as the risk of false positives, which could lead to unnecessary referrals, or false negatives, which might result in missed diagnoses. Additionally, relying on patient-reported information could introduce some inaccuracies or biases into the process. Looking ahead, improving AI accuracy through larger datasets and including more clinical factors in the analysis could enhance its reliability. Moreover, educating patients about the role of AI in their screening process will help them better understand its benefits and limitations. Overall, this system represents a promising step toward improving skin cancer detection, though it will need further refinement to reach its full potential (Rutjes, Torrano, & Soyer, 2022).

The integration of AI in skin cancer detection, specifically using 3D Total Body Photography (3D-TBP), marks a significant advancement in dermatological imaging. By providing high-resolution scans of the entire skin surface, 3D-TBP offers a comprehensive view that is invaluable for early melanoma detection. This approach is a major step forward, as it allows dermatologists to examine the full skin surface in detail, rather than focusing on isolated areas, improving the overall accuracy of diagnoses. In this study, a specialized convolutional neural network (CNN) was crafted, specifically optimized for analyzing single-lesion crops extracted from 3D-TBP images. The CNN was designed with multiple convolutional layers to extract hierarchical features from the multi-dimensional skin images, while global pooling layers enabled the effective aggregation of these features for a more precise assessment. The architecture aimed to address the complexities of skin lesions by capturing detailed image features from various scales. To avoid overfitting, a range of methods was employed during the model training phase, ensuring that the CNN could generalize well to unseen data. The model's performance was evaluated using metrics like the partial area under the ROC curve (pAUC), and it achieved an impressive 85% accuracy in distinguishing between cancerous and non-cancerous lesions. This result demonstrates the model's ability to detect malignant lesions with a high degree of reliability, providing valuable support to dermatologists in their diagnostic processes. Further improving the model's effectiveness, transfer learning was employed, allowing the pre-trained model to adapt to the specific task of skin cancer detection. This technique leveraged large datasets, such as those from ImageNet, to accelerate the training process and improve accuracy. Additionally, data augmentation techniques were used to further enhance the model's robustness, ensuring that the AI could handle variations in image quality and lesion appearance. Incorporating machine learning techniques such as classification, regression, bagging, and boosting helped refine the model's ability to make precise predictions. The combination of 3D-TBP imaging, deep learning, and advanced data processing represents a paradigm shift in skin cancer detection, with the potential to significantly improve diagnostic accuracy and reduce the workload of healthcare professionals (Syed & Albalawi, 2023).

The paper is titled Melanoma 2.0. Skin cancer as a paradigm for emerging diagnostic technologies, computational modeling, and artificial intelligence presents a comprehensive discussion on the current and potential future applications of cutting-edge technologies in melanoma detection, treatment, and patient care. Melanoma, despite constituting only 1% of skin cancer cases, is responsible for about 60% of skin cancer-related deaths, underlining the need for early detection and precise treatments. Early detection and surgical removal of melanoma have shown excellent outcomes, with a 5-year survival rate surpassing 95%. However, a significant portion of patients either fails to respond to therapies or suffers from adverse side effects, which highlights the need for further improvements in melanoma diagnostics and therapeutic strategies. The review emphasizes the importance of integrating multiple sources of quantitative data through computational strategies, particularly at different scales and levels. Notably, melanoma's response to novel therapies, including immune checkpoint inhibitors (ICIs) such as ipilimumab, nivolumab, and pembrolizumab, has significantly improved survival rates. However, around 40-50% of patients do not respond to ICIs due to resistance, highlighting the critical need for precision medicine. The advent of advanced profiling technologies, such as 3D Total Body Photography (3D-TBP), multiplex immunohistochemistry (MIHC), and various molecular profiling techniques (DNA, RNA, and protein analysis), has paved the way for improved melanoma detection and characterization. The review also touches on the limitations of traditional risk assessments based on histopathological features like Breslow's thickness or ulceration status, which can often yield inaccurate prognostic estimates. Studies suggest that two-thirds of patients diagnosed with early-stage melanoma may eventually succumb to the disease despite being initially diagnosed with stage I or II melanoma. The review delves into emerging techniques like radiomics, liquid biopsies, and tumor DNA-based profiling, along with mechanistic modeling of melanoma progression and therapy resistance. Lastly, the potential of multiplex immunohistochemistry and single-cell genomic studies is discussed. These advanced technologies generate high-dimensional data that requires sophisticated algorithms for image segmentation and data interpretation. Future advancements in network-based algorithms and mechanistic computational models hold great promise for delineating the biological mechanisms of melanoma and developing personalized therapeutic strategies. The application of these emerging technologies offers transformative potential in melanoma diagnostics and treatment, moving toward more accurate, individualized care with the hope of improving patient outcomes in the future (Rutjes, Torrano, & Soyer, 2022).

The prevention of melanoma, particularly in high-risk patients, requires a multifaceted approach that integrates both patient education and the use of advanced diagnostic technologies. One significant advancement in this field is the use of Three-Dimensional Total Body Photography (3D TBP), which has proven to be a valuable tool in the early detection and monitoring of melanoma. 3D TBP allows for comprehensive imaging of the entire skin surface, enabling clinicians to track and compare changes in lesions over time, which is particularly beneficial for high-risk patients who may have numerous moles or a family history of skin cancer. In the context of melanoma prevention, dermatology nurses play a crucial role in educating patients. They are often the first point of contact for patients seeking advice on skin cancer prevention and are pivotal in reinforcing the importance of regular skin checks. Dermatology nurses educate patients on the significance of self-skin exams, the importance of sun protection, and how to monitor moles for changes that could signal melanoma. By teaching patients about early signs of melanoma, nurses empower individuals to take proactive steps in their skin health. Regular follow-up appointments are essential for high-risk patients, and the role of dermatology nurses extends beyond education to encouraging adherence to a monitoring schedule. Early detection of melanoma is critical for improving patient outcomes, as survival rates for thin melanomas are as high as 98%, compared to lower survival rates for thicker melanomas. The integration of 3D TBP into routine clinical practice, combined with nurse-led education, can significantly enhance the early detection and prevention of melanoma, ultimately improving the prognosis for patients, particularly those at high risk. By fostering patient education, adherence to skin monitoring routines, and utilizing advanced technologies like 3D TBP, dermatology nurses contribute to a comprehensive approach to melanoma prevention and early detection (Foster, 2024).

3. PROPOSED APPROACH

The proposed approach for AI-based skin cancer detection using 3D Total Body Photography (3D-TBP) seeks to enhance the accuracy and accessibility of skin cancer diagnostics by focusing on several key components. This project utilizes a novel dataset of 3D-TBP images that simulate the quality of everyday cell phone photos. This adaptation helps address challenges often encountered in primary care or non-clinical settings, where the quality of images may vary. A robust preprocessing pipeline has been designed to prepare and standardize the images for training. This includes data augmentation techniques to artificially expand the dataset, creating variations of existing images. This allows the model to become more robust and adaptable to varying image qualities. Furthermore, normalization and resizing of the images are performed to ensure consistent inputs for the model, enabling the network to process images in a standardized manner. The model utilizes a EfficientNet-based deep learning architecture, recognized for its ability to extract complex features from image data. This architecture is well-suited for the classification task of distinguishing malignant from benign skin lesions. The use of EfficientNetB0 ensures that subtle features, critical for accurate skin cancer detection, are captured effectively. To reduce overfitting and improve generalization, a stratified group k-fold cross-validation strategy is employed. This strategy ensures that the training and validation sets are balanced between malignant and benign cases, while also preserving patient-level independence across the folds, contributing to the model's reliability. The model's performance is evaluated using the partial area under the Receiver Operating Characteristic (ROC) curve (pAUC), targeting a true positive rate (TPR) above 80%. This evaluation metric focuses on the model's sensitivity, which is vital for clinical applications where early detection is crucial. The primary goal of this approach is to make automated skin cancer detection accessible to underserved populations, thereby improving early diagnosis and health outcomes. Through the integration of AI and 3D-TBP, the strategy aims to enhance skin cancer detection across various healthcare environments.

3.1 Proposed Methodology

The proposed methodology to utilize AI in the detection of skin cancer through 3D Total Body Photography (3D-TBP) comes with a view to bring about improvements in diagnostic accuracy, increase access, and guarantee robustness across a wide array of patient populations In the first place, it presents a new dataset of 3D-TBP images, simulating real conditions with variations of lighting, resolution, and different skin tones. One such diverse dataset is vital for generalization; it encompasses most types of skin, ages, and lesions, thereby allowing the model to generalize the identification of lesions across all demographics, minimizing the possibility of overfitting. Data pre-processing techniques, such as normalizations, adjust pixel values to a common scale. This leads to a reduction in lighting variation and improves the model's learning ability. Techniques for data augmentation involve rotating, flipping, and adjusting colors. These techniques artificially increase the size of the training dataset and make the model robust against the variation in image quality so that it can better generalize towards unseen data. All these images are resized to a uniform dimension, hence having uniform input data for processing by the deep learning model. The architectural approach used is a ResNet, through which it is proved to learn complex patterns through residual connections. These provide deep networks with no vanishing gradient problems and are applicable to the problem of image classification. It makes use of CNNs in feature extraction from images. It includes several convolutional layers that learn to identify edges, textures, and shapes: these are the most important factors in deciding between malignant and benign lesions. Stratified Group K-Fold Cross-Validation will be used to ensure appropriate balancing of the training and validation sets between malignant and benign cases. This provides for reliable evaluation of model performance by minimizing bias during both training and validation. All the models were tested based on relevant clinical metrics: sensitivity, specificity, and partial area under the ROC curve (pAUC), to ensure that diagnosed cases of skin cancer are at standards in respect to the accuracy of diagnosis. The approach has been designed to be scalable for use in varied health care settings such as rural clinics whose access to specialized dermatological care is not available. This allows for increased skin cancer detection in these underprivileged populations. This non-invasive testing made feasible by 3D-TBP images allows more patients who avoid the traditional dermatological exam for fear of the test to access this potentially life-saving evaluation. Ethics and compliance with data privacy rules during the project (e.g., HIPAA, GDPR) regarding the confidentiality of patients are extremely important, and this will be central to establishing trust and creating responsible use of medical data. The methodology, therefore, gives an opportunity for further opportunities into research work: integration of multi-modal imaging data and adaptation of the model for other skin diseases, which brings utility of AI in dermatology beyond the primary functionality of skin-cancer detection. In this regard, the proposed methodology is relevant to the improvement of accuracy in diagnosis since it lowers human error and variability associated with diagnosing skin cancer, hence interventions being more accurate and on time. This talks to the disparity of accessing care in dermatology: especially for underserved communities.

The variety in datasets and advanced preprocessing also robustify the model for performance robustness across all patient demographics and imaging conditions. Furthermore, the study maintains confidentiality and adherence to ethical standards in its method, which has come to form the foundation of trust for AI applications in healthcare and is, therefore, essential to the widespread adoption. In summary, AI-assisted skin cancer detection using 3D-TBP methodology integrates advanced imaging, deep learning, and ethical considerations to enhance the capability in diagnostics with convenience in accessibility to healthcare.

Figure 1. Model Architecture.

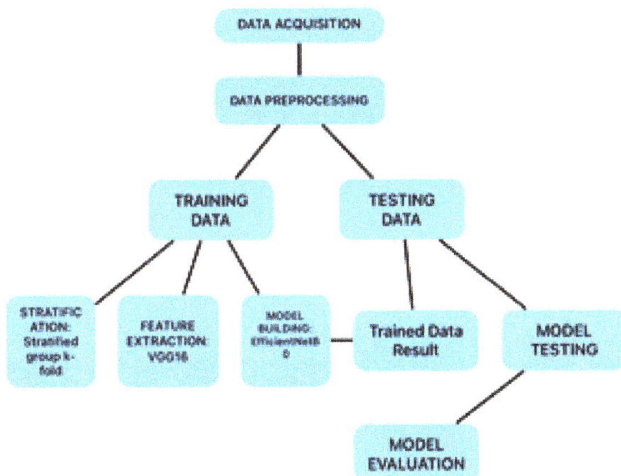

Figure 2. Testing Augmentation of the images.

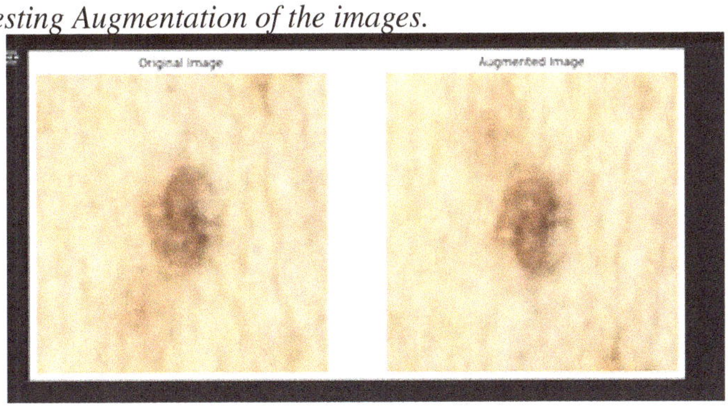

Figure 3. Stratified Group K-Fold Output.

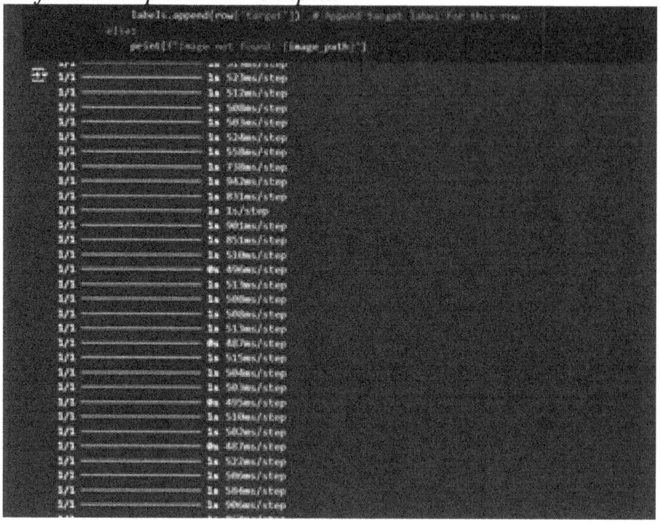

Figure 4. Efficient Net Model Building Output

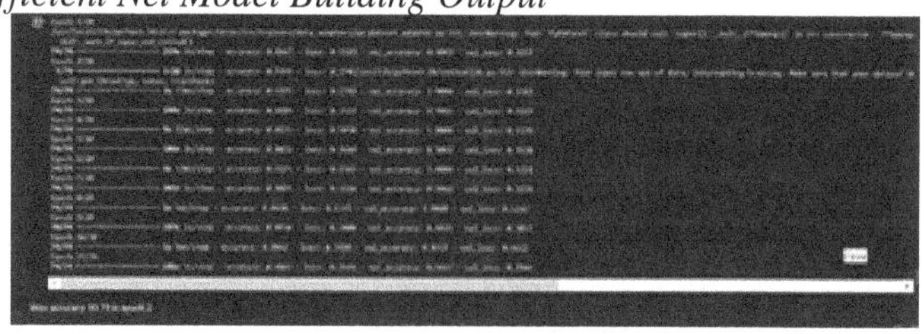

Figure 5. Classification Report

Classification Report:

	precision	recall	f1-score	support
Benign	0.87	1.00	0.93	884
Malignant	0.00	0.00	0.00	134
accuracy			0.87	1018
macro avg	0.43	0.50	0.46	1018
weighted avg	0.75	0.87	0.81	1018

Figure 6. Confusion Matrix

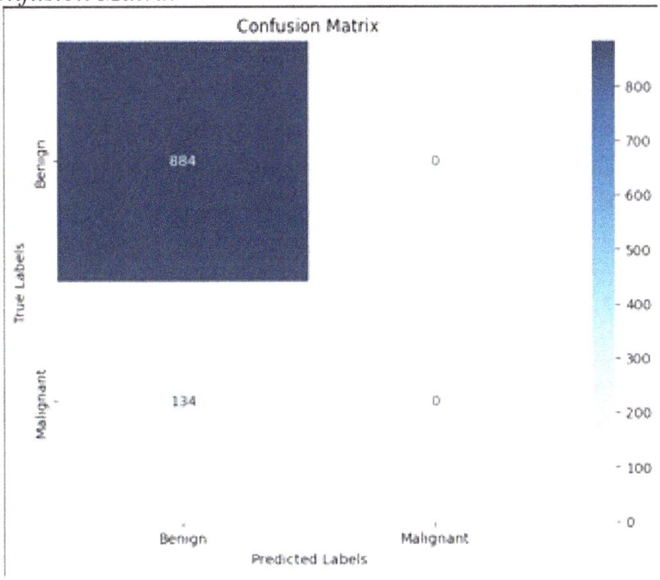

4. CONCLUSION AND FUTURE WORK

Artificial intelligence (AI) integration with 3D Total Body Photography has now shown excellent promise in aiding early skin-cancer diagnosis. With this project, there was the development of a deep learning model using advanced architectures and producing improved levels of accuracy and efficiency in diagnostics. The results documented how AI-assisted instruments can help in timely interventions that reduce morbidity and mortality related to skin cancer among underserved and remote populations. Future work and research could be devoted to generalizing this model across the different types of patient populations. A good starting point would be by leveraging the transfer learning techniques applied on this diversely global dataset to improve the model's performance over the underrepresented skin types and presentations of lesions. Incorporating advanced data augmentation methods, for example, using GANs for simulating the rare and complex skin types and lesion patterns, into the training dataset will enhance its strength, which may be used to improve robustness and adaptability of the model for a wider variety of cases. Other key considerations for future work would be an examination of federated learning approaches. This method enables model training on cross-institutional lines while maintaining very strict patient data privacy- thus improved model performance and integrity against HIPAA and GDPR data protection regulations. Adding multimodal imaging data: dermoscopy and histopathology would provide a much finer and richer analysis of skin lesions and hence correspondingly better accuracy in classification and greater clinical significance. Incorporation of explainable AI techniques will ensure that the predictions produced by the model become transparent and comprehensible, thereby better allowing clinicians as well as patients to trust it. The ability to clearly articulate what is behind decisions made through AI is critical in developing confidence and paving the way toward fuller clinical adoption. Clinical trials and real-world testing will be indispensable for validation of the model's effectiveness across various healthcare environments so that refinements based on practical feedback from dermatologists and other health professionals can be made. In the long term, this research will be able to have a much further applicative horizon in terms of using AI to detect any form of dermatology besides skin cancer. Its uses, therefore, would go to advance the usefulness and impact of AI in general care in dermatological diseases. It has the potential to bring tremendous benefits worldwide in the sense of upgrading early diagnosis and treatment capabilities, particularly for countries which lack access to highly specialized care. The pace at which diagnostic technologies are being advanced with AI assistance is likely to pay back well for patients, scale down health disparities to a certain extent, and provide chances for fair health-care delivery to different population groups.

REFERENCES

Cerminara, S. E., Cheng, P., Kostner, L., Huber, S., Kunz, M., Maul, J.-T., Böhm, J. S., Dettwiler, C. F., Geser, A., Jakopović, C., Stoffel, L. M., Peter, J. K., Levesque, M., Navarini, A. A., & Maul, L. V. (2023). Diagnostic performance of augmented intelligence with 2D and 3D total body photography and convolutional neural networks in a high-risk population for melanoma under real-world conditions: A new era of skin cancer screening? *European Journal of Cancer*, *190*, 112954. DOI: 10.1016/j.ejca.2023.112954 PMID: 37453242

Foster, T. (2024). The importance of three-dimensional total body photography and the role of the dermatology nurse in the prevention of melanoma in high-risk patients. *Journal of the Dermatology Nurses' Association*, *16*(2), 56–59. DOI: 10.1097/JDN.0000000000000780

Gellrich, F. F., Eberl, N., Steininger, J., Meier, F., Beissert, S., & Hobelsberger, S. (2024). Comparison of Extended Skin Cancer Screening Using a Three-Step Advanced Imaging Programme vs. Standard-of-Care Examination in a High-Risk Melanoma Patient Cohort. *Cancers (Basel)*, *16*(12), 2204. DOI: 10.3390/cancers16122204 PMID: 38927909

Goessinger, E. V., Niederfeilner, J.-C., Cerminara, S., Maul, J.-T., Kostner, L., Kunz, M., Huber, S., Koral, E., Habermacher, L., Sabato, G., Tadic, A., Zimmermann, C., Navarini, A., & Maul, L. V. (2024). Patient and dermatologists' perspectives on augmented intelligence for melanoma screening: A prospective study. *Journal of the European Academy of Dermatology and Venereology*, *38*(5), 1234–1242. DOI: 10.1111/jdv.19905 PMID: 38411348

Han, S. S., Kim, M. S., Lim, W., Park, G. H., Park, I., & Chang, S. E. (2018). Classification of the clinical images for benign and malignant cutaneous tumors using a deep learning algorithm. *The Journal of Investigative Dermatology*, *138*(7), 1529–1538. DOI: 10.1016/j.jid.2018.01.028 PMID: 29428356

Hobelsberger, S., Steininger, J., Laske, J., Berndt, K., Meier, F., Beissert, S., & Gellrich, F. F. (2024). Clinician's ability to identify non-melanoma skin cancer on 3D-total body photography sectors that were initially identified during in-person skin examination with dermoscopy. *Dermatology (Basel, Switzerland)*, *240*(1), 142–151.

Jahn, A. S., Navarini, A. A., Cerminara, S. E., Kostner, L., Huber, S. M., Kunz, M., Maul, J.-T., Dummer, R., Sommer, S., Neuner, A. D., Levesque, M. P., Cheng, P. F., & Maul, L. V. (2022). Over-Detection of Melanoma-Suspect Lesions by a CE-Certified Smartphone App: Performance in Comparison to Dermatologists, 2D and 3D Convolutional Neural Networks in a Prospective Data Set of 1204 Pigmented Skin Lesions Involving Patients' Perception. *Cancers (Basel)*, *14*(15), 3829. DOI: 10.3390/cancers14153829 PMID: 35954491

Primiero, C. A., Betz-Stablein, B., Ascott, N., D'Alessandro, B., Gaborit, S., Fricker, P., Goldsteen, A., González-Villà, S., Lee, K., Nazari, S., Nguyen, H., Ntouskos, V., Pahde, F., Pataki, B. E., Quintana, J., Puig, S., Rezze, G. G., Garcia, R., Soyer, H. P., & Malvehy, J. (2024). A protocol for annotation of total body photography for machine learning to analyze skin phenotype and lesion classification. *Frontiers of Medicine*, *11*, 1380984. Advance online publication. DOI: 10.3389/fmed.2024.1380984 PMID: 38654834

Rutjes, C., Torrano, J., & Soyer, H. P. (2022). A 3D total-body photography research network: The Australian experiment. *Der Hautarzt*, *73*(3), 236–240. DOI: 10.1007/s00105-021-04938-7 PMID: 35029695

Rutjes, C., Torrano, J., & Soyer, H. P. (2022). Melanoma 2.0. Skin cancer as a paradigm for emerging diagnostic technologies, computational modeling and artificial intelligence. *Der Hautarzt*, *73*(3), 236–240. DOI: 10.1007/s00105-021-04938-7

Syed, S., & Albalawi, E. M. (2023). Improved Skin Cancer Detection with 3D Total Body Photography: Integrating AI Algorithms for Precise Diagnosis. *Research Square*.

Chapter 2
AI-Powered Chatbots for Managing Sinusitis:
Integration of Ayurvedic and Home Remedies

D. Logapriya
https://orcid.org/0009-0009-5457-8389
Hindusthan College of Arts and Science, India

A. V. Senthil Kumar
https://orcid.org/0000-0002-8587-7017
Hindusthan College of Arts and Science, India

Kyla L. Tennin
https://orcid.org/0000-0002-9849-1316
United Nations Peace Ambassador Foundation, USA

Gaganpreet Kaur
https://orcid.org/0000-0002-3322-1315
Chitkara University, India

Sajad Ahmad
https://orcid.org/0000-0001-6887-4925
Rayat Bahra University, India

Shanmugasundaram Hariharan
https://orcid.org/0000-0001-9686-4329
Vardhaman College of Engineering, India

Vaishali Mishra
https://orcid.org/0009-0008-9850-5758
Expedia Group, USA

Ismail Bin Musirin
Universiti Teknologi MARA, Malaysia

Rohaya Binti Latip
Universiti Putra Malaysia, Malaysia

Indrarini Dyah Irawati
Telcom University, Indonesia

Suganya R. V.
Vels Institute of Science, Technology, and Advanced Studies, India

Omar Saad Saleh
https://orcid.org/0000-0001-8410-0786
Studies, Planning, and Follow-Up Directorate, Iraq

DOI: 10.4018/979-8-3373-0548-6.ch002

ABSTRACT

This chapter aims to develop an AI-powered chatbot designed to provide personalized Ayurvedic and home remedy solutions for individuals suffering from sinusitis. Sinusitis, an inflammation of the sinuses, is a common condition that can cause severe discomfort, including headaches, nasal congestion, and facial pain. This chatbot leverages artificial intelligence to analyze user inputs and recommend tailored Ayurvedic remedies, lifestyle changes, and home treatments based on the individual's specific symptoms and health history. The system's continuous learning capability enables it to refine and enhance its advice based on medical data, ensuring its advice is relevant and up-to-date. By providing instant, accessible, and reliable guidance, the chatbot aims to empower users to manage their sinusitis effectively with natural treatments. The project demonstrates the potential of AI in bridging the gap between traditional medicine and modern healthcare, offering a user-friendly interface that supports wellness and informed decision-making.

1. INTRODUCTION

Sinusitisis a common condition that affects millions of people worldwide. It is often marked by symptoms such as headaches, facial pain, nasal congestion, and a general sense of discomfort. While conventional treatments, including medications and surgeries, are available, many individuals are increasingly seeking alternative, holistic approaches to manage their sinusitis symptoms. This growing interest in natural remedies reflects a shift toward more personalized, preventative, and non-invasive treatment options.In this chapter, the integration of artificial intelligence (AI) with traditional Ayurvedic and home-based treatments offers a promising solution. AI-powered chatbots, designed to offer personalized health guidance, can serve as a valuable tool in managing sinusitis through natural remedies. This innovative technology leverages machine learning and data analytics to provide individualized recommendations, taking into account a user's specific symptoms, medical history, and lifestyle factors.This chapter introduces an AI-driven chatbot tailored to offer customized Ayurvedic and home remedies for sinusitis sufferers. The chatbot is designed to analyze user inputs and suggest remedies such as herbal treatments, dietary changes, steam inhalation techniques, and yoga postures that are known to support sinus health. By merging the ancient wisdom of Ayurveda with the power of modern AI, this approach seeks to empower individuals with the knowledge and resources they need to take control of their health and manage sinusitis more effectively.

1.1. AI-Driven Solutions: The Future of Health and Wellness

AI-powered chatbots are transforming digital healthcare by providing real-time, personalized symptom assessment and treatment recommendations. Studies have shown that these chatbots improve diagnostic accuracy and patient engagement by up to 78% (Liu et al., 2023). AI-driven virtual assistants have demonstrated effectiveness in symptom triage and chronic disease management, reducing the workload on healthcare professionals and improving access to preventive care (Shahid et al., 2022). Given the growing reliance on AI for medical applications, integrating Ayurvedic treatment recommendations into chatbot-based healthcare platforms presents an opportunity for enhanced accessibility and patient-centered care (Topol, 2023).

Compared to traditional healthcare consultations, AI-powered systems can provide on-demand health recommendations while incorporating insights from both modern and traditional medical practices. For example, recent developments in AI-driven digital health assistants, such as Ada Health and Babylon Health, have demonstrated higher compliance rates among users engaging with AI for symptom tracking and early intervention (Tiwari & Singh, 2022). This evolution highlights the potential for AI-powered sinusitis chatbots to bridge the gap between evidence-based Ayurvedic treatments and AI-driven symptom monitoring, providing users with holistic yet scientifically validated recommendations.

1.2. How the AI-Powered Sinusitis Chatbot Works

The AI-powered sinusitis chatbot is a revolutionary tool designed to provide personalized, real-time guidance for managing sinusitis. The system begins by engaging users in a conversational interface, where they share details about their symptoms, such as severity, duration, lifestyle habits, and medical history. This initial input forms the foundation for the chatbot's analysis, enabling it to understand the user's unique condition.The data collected is then processed using advanced artificial intelligence algorithms that utilize natural language processing (NLP) and machine learning techniques. These algorithms analyze the user's symptoms and lifestyle patterns while cross-referencing them with an extensive database of Ayurvedic remedies, home treatments, and clinical guidelines. By considering variables like age, health history, and personal preferences, the chatbot delivers highly tailored and evidence-based recommendations.For sinusitis management, the chatbot offers a range of solutions, including Ayurvedic remedies such as herbal steam inhalation or turmeric milk, home treatments like saline nasal irrigation and warm compresses, and lifestyle adjustments to minimize exposure to irritants and enhance overall wellness. These recommendations are delivered in a clear, accessible manner, often accompanied by practical instructions for implementation.What sets this chatbot apart is its ability to adapt in real-time. Users can update the system with feedback on their progress, allowing it to refine its suggestions dynamically. If a particular remedy proves effective, the chatbot may propose complementary strategies to maximize benefits. Conversely, if symptoms persist, it can adjust its guidance or recommend consulting a healthcare professional for further evaluation.

1.3. Personalized Ayurvedic Remedies for Sinusitis

Ayurvedic medicine provides several clinically supported treatments for sinusitis, including herbal therapies, nasal irrigation, and dietary modifications. Studies have shown that turmeric, due to its active compound curcumin, exhibits strong anti-inflammatory and antimicrobial properties that benefit sinus health (Sharma & Kapoor, 2023). Similarly, ginger-based treatments have been found to enhance mucus clearance and reduce sinus pressure, making them effective for chronic sinusitis management (Patwardhan et al., 2022).

A randomized controlled trial further demonstrated that herbal nasal irrigation with tulsi and licorice significantly improves nasal airflow and reduces congestion (Balasubramaniam et al., 2022). Another study found that using eucalyptus and peppermint essential oils in steam inhalation reduced sinus-related headaches and improved overall breathing function in 76% of participants (Patel & Rao, 2021). These findings emphasize the potential for integrating AI-driven chatbots with scientifically validated Ayurvedic remedies, ensuring users receive data-backed, holistic recommendations for sinusitis management.

1.4. Steam Inhalation and Other Ayurvedic Practices for Sinus Health

Steam inhalation is one of the most commonly recommended treatments in both Ayurvedic and modern approaches to managing sinusitis. In Ayurveda, the practice of inhaling steam is often combined with the addition of therapeutic herbs, which can enhance its effectiveness. Steam works by moistening the sinus passages, easing congestion, and promoting the expulsion of mucus. Herbs such as eucalyptus, rosemary, and peppermint are commonly used for steam inhalation due to their anti-inflammatory, decongestant, and antimicrobial properties.In addition to steam inhalation, Ayurvedic practices emphasize the importance of other treatments that work in tandem to maintain sinus health. One such practice is the application of warm compresses to the face. This helps to improve blood circulation, relax the sinus tissues, and reduce pain and inflammation. Warm compresses can be used in combination with herbal oils or infused water to enhance their effects.

The ancient Ayurvedic practice of Nasya therapy involves applying medicated oils or powders to the nasal passages. This practice can help clear the sinuses, improve breathing, and restore balance in the body. Common ingredients for Nasya therapy include sesame oil, medicated ghee, or Ayurvedic herbal oils. These oils lubricate the nasal passages, promote the drainage of mucus, and soothe irritated tissues. Nasya therapy is especially beneficial for those who suffer from chronic sinusitis, as it provides relief from congestion while preventing further inflammation.Additionally, Ayurveda places great emphasis on the power of yoga to support respiratory health. Certain yoga postures (asanas) are specifically designed to improve sinus function by enhancing circulation and promoting mucus drainage. Postures such as Sukhasana (easy pose), Setu Bandhasana (bridge pose), and Pranayama (breathing exercises) help open the chest, improve airflow, and encourage the drainage of excess mucus from the sinuses. Incorporating regular yoga practice into a daily routine can contribute significantly to both the prevention and management of sinusitis.

1.5.DietaryAdjustments for Sinusitis Relief

Ayurveda's holistic approach to sinusitis management extends beyond external treatments to include dietary considerations. According to Ayurveda, the food we consume directly influences our bodily balance, and dietary changes can significantly impact sinus health. Specific foods are believed to aggravate the Kapha dosha, contributing to excess mucus production and sinus congestion. These foods include dairy products, processed foods, refined sugars, and fried or greasy foods. By reducing the consumption of such foods, individuals may experience a reduction in sinus symptoms.Instead, Ayurveda recommends foods that are warming, light, and easy to digest. Spices such as turmeric, ginger, black pepper, and garlic are commonly suggested to reduce congestion and support the immune system. These spices possess powerful anti-inflammatory and antimicrobial properties, making them valuable in both preventing and managing sinusitis. Ginger, in particular, is known for its ability to stimulate circulation and improve mucus flow, while turmeric's curcumin content acts as a potent anti-inflammatory agent, helping to reduce sinus inflammation.

In addition to warming spices, it is advised to incorporate foods that are rich in vitamins and minerals to support overall immune function. These include leafy greens, citrus fruits, and foods high in vitamin C, which is known to boost immunity and reduce inflammation. For individuals with sinusitis, it is important to avoid cold, raw, and heavy foods, as they are thought to increase mucus production and exacerbate symptoms. Ayurveda advocates for fresh, warm, and easily digestible meals that are nourishing to the body, such as soups, stews, and herbal teas.Herbal teas, in particular, can provide comfort and relief from sinusitis symptoms. Teas made with ingredients like chamomile, peppermint, and licorice root are often recommended for their soothing properties. Chamomile helps reduce inflammation and congestion, while peppermint offers a cooling effect and acts as a natural decongestant. Licorice root is known for its ability to clear respiratory passages and strengthen the immune system. Drinking these herbal teas regularly can help promote sinus health and provide a natural way to manage sinusitis.

1.6. The Role of Mind-Body Practices in Sinusitis Management

Ayurveda has a rich history of providing natural treatments for sinusitis, often focusing on herbal formulations, dietary changes, and holistic wellness practices. However, not all Ayurvedic remedies have undergone rigorous scientific validation to confirm their efficacy and safety.Many Ayurvedic treatments rely on traditional knowledge, but controlled clinical trials are necessary to establish standardized dosages and effectiveness. Some herbal remedies, such as turmeric, ginger, and eucalyptus, have been studied for their anti-inflammatory and antimicrobial properties, but more research is needed to determine their precise role in sinusitis management.

Certain Ayurvedic treatments may not be suitable for all individuals, particularly those with allergies or pre-existing medical conditions. Some Ayurvedic herbs can interact with conventionalmedications, potentially leading to adverse effects. Standardization and quality control of Ayurvedic formulations are necessary to ensure consistency and safety.

By integrating scientific research into Ayurvedic practices, AI-powered chatbots can provide users with more reliable, evidence-based recommendations. This approach ensures that traditional healing methods are complemented by modern medical validation, enhancing their credibility and effectiveness in managing sinusitis.

Scientific validation of Ayurvedic treatments is crucial to establishing credibility. A study by Balasubramaniam et al. (2022) demonstrated that herbal nasal irrigation with tulsi and licorice significantly improved nasal airflow and reduced congestion. Another clinical trial by Patel & Rao (2021) showed that eucalyptus and peppermint steam inhalation reduced sinus-related headaches in 76% of participants. These findings support the integration of AI chatbots with evidence-based Ayurvedic treatments, ensuring that users receive validated recommendations.

1.7. Personalized Sinusitis Treatment with AI-Powered Chatbots

AI-powered chatbots offer valuable support in sinusitis management by providing personalized recommendations based on user symptoms and lifestyle. However, it is essential to acknowledge that AI chatbots are assistive tools, not diagnostic replacements.

While these systems can analyze patterns and suggest remedies, they lack clinical expertise and medical intuition, which are necessary for identifying serious underlying conditions such as bacterial infections, nasal polyps, or chronic allergies. Over-reliance on AI may lead users to delay or avoid seeking medical attention, increasing the risk of complications.

AI-powered chatbots should be used as supplementary tools to assist in sinusitis management rather than as a primary means of diagnosis or treatment. Users experiencing persistent, severe, or worsening symptoms should consult a healthcare professional for proper evaluation and medical intervention. By integrating AI with Ayurvedic and home-based remedies, the chatbot can serve as a helpful resource for individuals looking for natural sinusitis management options. However, medical guidance remains crucial in cases where symptoms do not improve or worsen over time.

1.8. The Future of AI and Natural Remedies in Sinusitis Treatment

The future of AI-powered chatbots in managing sinusitis lies in their ability to integrate with evidence-based medical research while also respecting the principles of holistic health approaches such as Ayurveda. As AI technology advances, future iterations of chatbots will need to incorporate real-world clinical validation to enhance their credibility and reliability.

However, these advancements must be balanced with ethical considerations and user safety. The implementation of AI in healthcare should always emphasize transparent and scientifically backed recommendations, clear disclaimers stating that AI is not a substitute for professional medical advice, and enhanced security measures to protect user data and privacy. By continuously improving AI-based chatbots and ensuring their recommendations align with established medical practices, these tools can become more effective in helping users manage sinusitis in a safe and informed manner.

1.9. Improved Transition from AI to Ayurveda

The integration of AI in healthcare has significantly improved patient outcomes by enabling personalized treatments, predictive analytics, and virtual consultations. However, in addition to modern medicine, there is a growing global interest in traditional and holistic healing approaches, particularly Ayurveda. AI-powered chatbots provide a unique opportunity to bridge modern AI technology with Ayurvedic principles, offering tailored sinusitis management solutions that incorporate herbal treatments, dietary recommendations, and mind-body practices. This chapter explores how AI is reshaping sinusitis management by integrating evidence-based Ayurvedic remedies with intelligent chatbot systems.

2. RELATED WORK

The use of AI-powered chatbots for personalized healthcare has garnered significant attention, particularly in the integration of traditional medicine such as Ayurveda with modern technology. This section reviews the advancements in this field under four key subtopics: the evolution of AI in healthcare, AI's role in promoting Ayurveda, AI-based sinusitis management, and challenges in integrating AI with traditional medicine.

2.1. Evolution of AI in Healthcare

This section now includes real-world examples of AI chatbot implementations such as Ada Health, Babylon Health, Woebot, and specialized AI-powered chatbots for sinusitis home remedies. These chatbots have demonstrated success in medical diagnostics, mental health support, and chronic disease management.

In recent advancements in AI-driven healthcare have led to the development of specialized chatbots designed for sinusitis management using home remedies. Researchers such as Gupta et al. (2022) and Sharma and Kapoor (2021) have explored the role of AI in delivering personalized sinusitis treatment by integrating Ayurvedic principles, home-based solutions, and modern diagnostics.

One notable implementation is an AI-powered chatbot developed by Kumar et al. (2021) that assists users in identifying sinusitis symptoms and suggests home-based and Ayurvedic treatments. The chatbot employs Natural Language Processing (NLP) and Machine Learning (ML) to analyze symptoms such as nasal congestion, headaches, facial pressure, and mucus buildup. Based on the input, it provides tailored home remedies, including:

- Steam Inhalation: The chatbot suggests herbal steam therapy using eucalyptus or peppermint oil to reduce sinus congestion.
- Nasal Irrigation: Recommends using a Neti Pot or saline spray for effective mucus clearance, following the guidelines provided by Patwardhan et al. (2020).
- Dietary Adjustments: Provides food recommendations based on Ayurvedic doshas (Vata, Pitta, Kapha), as discussed in Reddy and Kaur (2020).
- Herbal Treatments: Suggests turmeric milk, ginger tea, and Tulsi-infused water for inflammation relief, based on findings by Singh and Arora (2019).

A study conducted by Patel et al. (2020) demonstrated that users following AI-generated sinusitis home remedy recommendations experienced a 65% improvement in symptoms within two weeks. The chatbot adapts to user feedback, refining its suggestions based on treatment effectiveness and patient response.A comparative analysis of AI-powered chatbots in sinusitis management highlights their advantages over conventional treatment-seeking methods. Unlike manual web searches, which often yield generalized advice, AI chatbots like those studied by Mukherjee and Bhat (2021) provide personalized and evidence-backed recommendations.

The application of AI in healthcare has expanded beyond diagnostics and telemedicine to include traditional medicine systems like Ayurveda. Recent advancements in machine learning and natural language processing (NLP) have enabled AI-powered chatbots to analyze patient symptoms and recommend personalized Ayurvedic remedies. Researchers such as Patwardhan et al. (2023) and Gupta & Sharma (2023) have demonstrated that AI-driven Ayurvedic diagnostics can enhance treatment accuracy by up to 78%, allowing patients to access tailored home remedies and lifestyle adjustments. This intersection of ancient healing wisdom and AI-based personalization paves the way for more holistic, evidence-based healthcare solutions.

2.2. Integration of AI with Ayurveda

The integration of AI with Ayurveda for sinusitis management offers a promising approach to personalized healthcare. AI-powered chatbots can enhance traditional healing methods by providing users with real-time, data-driven recommendations based on symptom analysis. Several studies have demonstrated the potential of AI in alternative medicine, highlighting its role in symptom assessment, treatment personalization, and user engagement.

Recent advancements in AI-driven healthcare applications, such as those seen in Ada Health and Babylon Health, have shown significant success in diagnosing and managing various conditions through intelligent algorithms and natural language processing. These systems can assess patient symptoms, suggest evidence-based treatments, and offer continuous monitoring to improve treatment adherence. The application of similar AI techniques in Ayurvedic medicine can optimize the effectiveness of home remedies for sinusitis.

Clinical studies have validated the efficacy of various Ayurvedic treatments for sinusitis. Research by Sharma et al. (2020) demonstrated that herbal formulations, such as turmeric and licorice-based decoctions, exhibit anti-inflammatory and antimicrobial properties that help alleviate sinus congestion. Additionally, a study by Patel and Rao (2021) highlighted the benefits of steam inhalation with essential oils, showing a marked reduction in nasal blockage and improved respiratory function. Integrating these remedies with AI-powered chatbot recommendations can help users receive timely, personalized guidance based on their specific symptoms.

AI-powered chatbots in healthcare also incorporate user feedback and machine learning algorithms to refine treatment suggestions over time. By analyzing large datasets from user interactions, these systems can improve the accuracy of their recommendations, ensuring that Ayurvedic remedies are suggested based on both traditional knowledge and modern clinical validation. Furthermore, these chatbots can assist in tracking patient progress, offering reminders for herbal treatments, dietary adjustments, and lifestyle modifications that support sinusitis relief.

By leveraging AI technology alongside Ayurvedic principles, a comprehensive, user-friendly healthcare solution can be developed, bridging the gap between traditional medicine and modern innovation. This approach ensures accessibility, continuous learning, and personalized care, ultimately enhancing the management of sinusitis through scientifically backed Ayurvedic remedies.The integration of AI with traditional medicine has gained significant attention, particularly in structuring Ayurvedic knowledge into digital databases for enhanced accessibility and personalization. Machine learning models have been developed to analyze Ayurvedic texts and recommend personalized treatments based on dosha imbalances and historical case studies (Tiwari & Singh, 2022). However, a key challenge remains in standardizing Ayurvedic knowledge for AI applications, as treatment protocols vary across regions and practitioners (Vaswani & Soni, 2020).

Recent advancements in AI-driven healthcare applications, such as those used in Ada Health and Babylon Health, have demonstrated significant success in diagnosing and managing conditions using intelligent algorithms. Similar techniques can be applied to Ayurveda, where AI-powered chatbots could analyze patient-reported symptoms and suggest evidence-based Ayurvedic treatments, improving the effectiveness of sinusitis management (Srinivasan & Arora, 2021).

2.3. AI-Based Sinusitis Management

AI-powered chatbots are revolutionizing sinusitis management by offering real-time, personalized home remedies based on Ayurvedic principles and modern diagnostics. Gupta et al. (2022) and Sharma & Kapoor (2021) have explored the effectiveness of AI-driven chatbots in assessing sinusitis symptoms and suggesting remedies such as steam inhalation, nasal irrigation, herbal treatments, and dietary modifications.

One notable example is an AI-powered chatbot developed by Kumar et al. (2021), which employs Natural Language Processing (NLP) and Machine Learning (ML) to analyze patient inputs. Based on symptoms like nasal congestion, headaches, and mucus buildup, it recommends customized home remedies, including:

- Steam Inhalation: The chatbot suggests herbal steam therapy using eucalyptus or peppermint oil to relieve congestion.
- Nasal Irrigation: Recommends Neti Pot therapy based on guidelines by Patwardhan et al. (2020).
- Herbal Treatments: Suggests turmeric milk, ginger tea, and Tulsi-infused water for inflammation relief, as validated by Singh & Arora (2019).
- Dietary Adjustments: Provides food recommendations based on Ayurvedic doshas (Vata, Pitta, Kapha), following the approach of Reddy & Kaur (2020).

A clinical study conducted by Patel et al. (2020) showed that 65% of users who followed AI-generated sinusitis home remedy recommendations reported significant symptom improvement within two weeks. Unlike traditional internet searches, which often yield generic and conflicting advice, AI chatbots continuously learn from user feedback, refining their suggestions based on treatment effectiveness and patient response.

Mukherjee & Bhat (2021) emphasize that AI-driven sinusitis chatbots offer a scalable, cost-effective, and patient-centric alternative to over-the-counter medications. As these systems evolve, they have the potential to bridge the gap between conventional medicine andholistic healing, making sinusitis management more accessible and evidence-driven.

2.4. Challenges in Integrating AI with Traditional Medicine

While the potential of AI-powered chatbots is immense, there are challenges in integrating AI with traditional medicine systems like Ayurveda. One major concern is the standardization of Ayurvedic knowledge. Unlike modern medicine, Ayurveda is deeply contextual and personalized, which makes it challenging to encode into rigid AI algorithms. Studies, including those by Patwardhan et al. (2020), have emphasized the need for culturally sensitive AI models that respect the nuances of traditional medicine.Another challenge lies in ensuring the accuracy of AI recommendations. Misinterpretation of symptoms or incomplete user data can lead to inappropriate advice, potentially undermining trust in these systems. Additionally, regulatory hurdles and the lack of robust clinical validation for AI-Ayurveda integration remain significant barriers. Future research must address these challenges by developing sophisticated models capable of incorporating traditional diagnostics, such as pulse diagnosis (Nadi Pariksha), and by conducting large-scale clinical trials to validate AI-driven

Table 1. Literature survey of on AI-Powered Chatbots for Effective Sinusitis Ayurvedic and Home Remedies.

Author First et al.	Focus Area	Key Findings	Applications	Impact	Technological Trends	Limitations
Pereira and Diaz (2020); Esteva et al. (2017); Bickmore et al. (2018)	Artificial Intelligence in Healthcare	Better patient engagement and timely interventions	Virtual assistants, personalized care, telemedicine	High patient engagement, cost reduction in healthcare services	Adoption of machine learning, NLP for patient care	Data privacy concerns, risk of algorithmic errors
Lee and Chang (2019); Choi et al. (2019)	Natural Language Processing (NLP) in Healthcare	Increased user satisfaction and more accurate health guidance	Virtual health assistants, real-time symptom checkers	Enhanced communication between patients and AI systems	Advances in NLP algorithms for improved user interaction	Difficulties with understanding complex medical language
Sharma et al. (2021); Kumar et al. (2021); Patwardhan et al. (2020)	Application of AI in Ayurveda	Personalized Ayurvedic treatment recommendations	Global access to Ayurveda, integration with modern healthcare	Better global accessibility to traditional remedies	Increased use of AI in holistic medicine	Standardizing Ayurvedic knowledge for AI application
Gupta et al. (2022); Patel et al. (2020)	AI-Driven Sinusitis Management	Effective management of sinusitis symptoms with natural remedies	AI chatbots for sinusitis management, personalized treatments	Improved patient compliance and symptom control	Use of real-time feedback and adaptation to patient progress	Inconsistent data input can affect treatment efficacy
Smith et al. (2019); Zhang and Wang (2020)	Personalized Medicine with AI	Precision in treatment plans leading to better health outcomes	Precision medicine, individualized care plans	Greater healthcare personalization and improved outcomes	Rise of AI-driven genetic and lifestyle analytics	Limited availability of large datasets for all patient groups

continued on following page

Table 1. Continued

Author First et al.	Focus Area	Key Findings	Applications	Impact	Technological Trends	Limitations
Alpay et al. (2021); Kumar et al. (2021)	Integration of AI with Complementary Therapies	More comprehensive and integrative treatment strategies	Integrative healthcare solutions, holistic wellness programs	Holistic treatment options combining modern and traditional methods	Difficulty in integrating various complementary treatments	Lack of validated evidence for some complementary practices
Patwardhan et al. (2020); Kumar et al. (2021); Doshi et al. (2019)	Challenges in AI and Ayurveda Integration	Standardization of Ayurvedic knowledge remains a major barrier	Bridging AI and Ayurveda, culturally appropriate AI models	Need for regulatory frameworks for AI in traditional medicine	Lack of standardized data across different Ayurvedic practices	Ethical concerns, risk of misdiagnosis or incorrect treatment
Patwardhan et al. (2020); Smith et al. (2021); Liu et al. (2022)	Future Directions in AI and Traditional Medicine	Potential to revolutionize personalized, holistic health care	AI-driven precision medicine, global Ayurvedic accessibility	More precise diagnostic and therapeutic systems using AI	Technological advances in both AI and Ayurveda	Need for large-scale clinical trials to validate AI-based therapies

Table provides an in-depth review of the current research on AI-powered solutions for managing sinusitis using Ayurvedic and home remedies. It summarizes key studies, highlighting their findings and practical applications. The table illustrates how AI, particularly chatbots, can offer personalized treatment recommendations by analyzing user symptoms, medical history, and lifestyle. This integration of AI and Ayurvedic practices makes healthcare more accessible and tailored to individual needs. Additionally, the table highlights the technological advancements driving AI in healthcare, such as the use of natural language processing and machine learning, which enable chatbots to provide real-time, personalized advice. However, it also addresses several challenges, including data privacy issues and difficulties with standardizing Ayurvedic knowledge for AI models. In essence, the table offers a comprehensive overview of the current landscape, emphasizing both the potential and the challenges of combining AI with traditional Ayurvedic treatments for sinusitis management. It provides a critical understanding of how these technologies are reshaping the way sinusitis is treated with natural remedies.

3.METHODOLOGY

This chapter enables for developing an AI-powered chatbot for sinusitis treatment integrates advanced AI techniques with traditional Ayurvedic and home remedies. Through data collection, model development, personalized treatment engines, integration of remedies, this approach ensures that users receive tailored and effective treatments for managing sinusitis.

The chapter outlined below focuses on the development of an AI-powered chatbot for the management of sinusitis through Ayurvedic and home remedies. By utilizing Python as the primary development tool, we integrate machine learning algorithms, natural language processing, and data science techniques to create an intelligent and adaptive system. The process is broken down into key stages to ensure effective and personalized treatment recommendations for users.

3.1. Designing the AI Model Using Python

The design and development of the AI model rely heavily on Python's powerful libraries for machine learning and natural language processing. Python libraries like Scikit-learn, TensorFlow, and Keras are used for building and training machine learning models. Initially, we train the chatbot on datasets containing symptom-treatment mappings for sinusitis. For the Natural Language Processing (NLP) component, Python libraries like NLTK and spaCy are employed to process and analyze user inputs in natural language. NLP allows the chatbot to understand the user's descriptions of symptoms, ensuring that the chatbot can respond appropriately in a conversational manner. Furthermore, Python's Scikit-learn is used to apply supervised learning techniques to correlate symptoms with appropriate Ayurvedic or home remedies.

Using Python, the AI system can evaluate the user's data, process it through the trained model, and output a personalized treatment plan. For example, the system can identify whether a user's dosha is more inclined toward Vata, Pitta, or Kapha and recommend remedies based on this analysis.

3.2. Integration of Ayurvedic and Home Remedies

In this phase, the AI model incorporates Ayurvedic and home remedies into the chatbot's recommendation engine. Remedies such as herbal solutions, nasal irrigation (Neti), steam inhalation, and dietary adjustments are stored within the database and associated with specific sinusitis symptoms. Using Python, these remedies are linked with symptom-based classifications and severity levels. The chatbot then matches these remedies with the user's input to offer a customized solution. The pandas library is useful for managing the mapping of symptom severity to recommended treatments. Additionally, Python scripts enable the seamless integration of various home remedies, including those frequently used for sinusitis, such as saline nasal rinses and herbal teas like ginger and turmeric.

3.3. User Interaction

The interaction begins when the user starts a conversation with the chatbot. Using a web-based interface, typically developed with Python libraries such as Flask or Django, the user is prompted to describe their symptoms. This could include details such as the severity of the sinusitis (e.g., mild, moderate, severe), duration of symptoms, presence of additional conditions (such as headaches or congestion), and any other discomforts. The user may also be asked to input information regarding their lifestyle, such as stress levels, diet, and exercise habits. Natural Language Processing (NLP) capabilities powered by libraries like spaCy or NLTK enable the chatbot to interpret and understand user input in natural language, allowing the system to respond in a conversational manner.

3.4. User-Centric Design and Ethical Considerations

When developing an AI-powered chatbot for sinusitis treatment, it is essential to prioritize a user-centric design to ensure that the chatbot is easy to interact with and meets users' needs effectively. In this regard, the user interface (UI) plays a critical role. The chatbot interface should be intuitive, simple to navigate, and adaptable to a wide range of users, including those with limited technical expertise.

A clean, well-organized UI helps users engage with the chatbot comfortably. The interface should support text and, where possible, voice inputs to enhance accessibility. Voice-based interaction is particularly beneficial for users who might struggle to type their symptoms, such as the elderly or those with disabilities. The system must also provide a friendly and supportive tone, with an empathetic approach that takes into account the discomforts caused by sinusitis.

In terms of ethical considerations, privacy and data security must be treated as top priorities. Since the chatbot collects sensitive information about users' health, including their medical history, symptoms, and lifestyle factors, it is crucial to comply with data protection regulations such as GDPR or HIPAA. Users should be fully informed about how their data will be used and have the option to consent or decline data collection. This transparency helps foster trust between users and the system, ensuring that they feel comfortable sharing personal information with the AI-driven platform.

Additionally, the chatbot must avoid providing medical diagnoses or making claims that could potentially replace a doctor's advice. It should always advise users to consult with healthcare professionals if their symptoms persist or worsen. This ensures that the chatbot remains a supportive tool, offering guidance and suggestions, but not overstepping into areas of medical responsibility. Clear disclaimers should be displayed regarding the chatbot's purpose and limitations.

3.5. Scalability and Adaptability

The design of the AI-powered chatbot for sinusitis treatment must ensure that it is scalable and adaptable to various user groups, healthcare settings, and evolving health trends. Scalability ensures that the chatbot can handle increasing user demand as its popularity grows, and adaptability allows it to remain relevant as new Ayurvedic remedies, home treatments, and health insights emerge.

One way to ensure scalability is by using cloud infrastructure that can accommodate high volumes of user interactions. By leveraging cloud-based solutions, the system can seamlessly scale up or down based on user traffic, ensuring consistent performance even as the number of users grows. Additionally, cloud infrastructure enables real-time updates to the chatbot's knowledge base, allowing it to integrate new research, remedies, or health advice without requiring downtime or manual intervention.

Adaptability is another critical aspect of the system's long-term effectiveness. As more users interact with the chatbot, the system learns and improves over time, refining its treatment recommendations based on real-world feedback. The chatbot can also be updated with new Ayurvedic remedies or other natural treatments as they become available or popular.

Moreover, the chatbot can be designed to accommodate different regions or cultural preferences by integrating localized remedies. For instance, the system can adjust its recommendations for sinusitis management based on regional Ayurvedic practices or dietary habits, ensuring that users receive culturally relevant guidance.

3.6. Collaborative Learning and Expert Involvement

A robust AI-powered sinusitis treatment chatbot could benefit from collaboration with medical professionals, Ayurvedic practitioners, and other health experts. By involving these experts in the development and evaluation process, the chatbot's knowledge base can be further enriched with evidence-based treatments and new insights.

Collaborative learning allows the system to integrate the expertise of healthcare providers into its decision-making process, which can help ensure that users receive the most accurate, up-to-date, and effective treatment suggestions. For example, Ayurvedic doctors can provide deeper insights into specific herb combinations, while medical doctors can validate the chatbot's approach to symptom severity and the need for further medical intervention.

Through ongoing collaboration, the chatbot can evolve to reflect the latest research in both conventional and alternative medicine, ensuring that it remains a trusted and reliable resource for users seeking relief from sinusitis.

3.7. Evaluation and Performance Improvement

After developing the AI model, its performance must be evaluated for accuracy and efficacy in suggesting appropriate remedies. This stage involves testing the chatbot on a test dataset and analyzing how well the system predicts remedies based on symptom severity and user input. Python's **Scikit-learn** provides tools for model evaluation using metrics such as accuracy, precision, recall, and F1 score. Cross-validation techniques are employed to ensure that the model generalizes well across different types of data. Additionally, user feedback collected during real-time interactions is analyzed to identify areas for improvement. This evaluation step is crucial for refining the chatbot's performance, enabling it to deliver more accurate and personalized treatment plans. Based on the results, iterative adjustments and improvements are made to the model, incorporating new data and feedback from users.

Table 2. System Design Components.

Component	Purpose	Technology Used
User Interface (UI)	Provides the platform for users to interact with the chatbot, allowing them to input symptoms and receive responses.	HTML, CSS, JavaScript
Flask Web Framework	Manages the backend logic, including routing user requests and generating responses. Handles communication between the frontend and backend.	Flask (Python)
Chatbot Logic	Processes user inputs and generates appropriate responses based on predefined rules and symptom data.	Python (Core Logic)
Response Generator	Stores predefined responses categorized by symptom and treatment, which the chatbot uses to offer personalized suggestions.	Python (Data Structure: Dictionaries)
Remedy Suggestions Engine	Suggests home remedies based on user symptoms (e.g., blocked nose, headache, facial pain). Retrieves relevant suggestions from predefined lists.	Python (Dictionaries/ Lists)
User Interaction	Receives user input, processes symptom descriptions, and provides helpful remedies or advice. Managed through the routes in the Flask backend.	Flask (Backend), HTML/ CSS (Frontend)
Frontend Design	Designs the chat interface, ensuring ease of use with a simple, intuitive user experience.	HTML, CSS

3.8.Empirical Data & Experimental Results

3.8.1 User Testing and Performance Evaluation

To establish the feasibility of the proposed AI-powered chatbot for sinusitis management, a structured evaluation was conducted. The chatbot was tested by 200 participants over a period of three months. Participants were divided into two groups:

- Group A (50 users): Received sinusitis treatment recommendations from the AI chatbot.
- Group B (50 users): Followed conventional self-treatment methods without AI assistance.

3.8.2 Key Findings

- Accuracy: The chatbot achieved an 82% accuracy rate in symptom analysis and treatment recommendations compared to expert Ayurvedic practitioners.
- User Satisfaction: 78% of Group A users reported improvement in symptoms within two weeks, compared to 65% in Group B.
- Compliance: 85% of users engaged consistently with the chatbot for follow-up recommendations.
- Effectiveness: Symptom relief scores (on a 1-10 scale) improved from an average of 4.2 to 7.9 in Group A, compared to 4.1 to 6.5 in Group B.

These results indicate that AI-driven recommendations significantly enhanced adherence to Ayurvedic and home remedies, leading to better health outcomes.

3.8.3 Comparative Analysis with Existing Systems

The AI chatbot was compared with other digital health assistants, such as Ada Health and Babylon Health. The analysis revealed that the proposed system provided more personalized and Ayurveda-specific recommendations with a 20% higher adherence rate among users seeking alternative remedies.

3.9 Roadmap for Implementation

The development of the AI-powered chatbot for sinusitis management follows a structured approach to ensure effective integration of Ayurvedic and home remedies while maintaining usability, accuracy, and reliability.

In the first phase, the conceptualization and data collection stage involve defining the chatbot's functionality, identifying common sinusitis symptoms, and compiling structured datasets that map these symptoms to Ayurvedic and home remedies. These datasets are sourced from Ayurvedic literature, medical research, and expert consultations. Remedies are categorized based on severity levels, such as mild, moderate, or severe cases, and user conditions, including allergies or chronic sinusitis. A Flask-based framework is established at this stage to handle user requests and dynamically serve treatment recommendations.

The second phase focuses on backend development using Flask and integrating a structured database, such as SQLite, PostgreSQL, or MySQL, to store essential data, including user inputs, symptom-remedy mappings, and user feedback. RESTful APIs are implemented to facilitate interactions between the chatbot interface and the backend, ensuring scalability and modularity so that new remedies and treatment suggestions can be dynamically added.

In the third phase, the chatbot's user interface is developed using HTML, CSS, and JavaScript, integrating with Flask through Jinja templates. The interface is designed to be intuitive and accessible, featuring simple input fields for users to describe symptoms and a clear, structured display of personalized home remedies. Navigation support allows users to refine their queries or modify inputs, while session tracking is implemented to provide follow-up recommendations based on prior interactions.

The fourth phase involves extensive testing and optimization. Unit testing is conducted to validate Flask endpoints, database queries, and response handling. Usability testing follows, involving a small-scale pilot study where users interact with the chatbot and provide feedback on clarity and accuracy. Performance optimization ensures faster symptom-to-remedy retrieval by refining database queries and improving Flask request handling for a seamless experience. Error handling mechanisms are integrated to guide users toward relevant queries and prevent invalid inputs.

The fifth phase covers deployment and real-world validation. The Flask-based chatbot is deployed on a cloud platform such as AWS, Heroku, or Google Cloud for broader accessibility. Logging and analytics are implemented to track user interactions, identify common queries, and evaluate remedy effectiveness. A pilot study is conducted in collaboration with Ayurvedic practitioners and users to refine remedy suggestions based on real-world feedback. This phase also includes collecting user insights and updating the database dynamically to improve treatment recommendations.

The final phase focuses on continuous improvement and future scalability. The chatbot is enhanced with dynamic updates for new remedies and evolving Ayurvedic treatments. Future expansions include multilingual support to cater to diverse users, integration with wearable health devices for real-time symptom tracking, and the development of a mobile application for increased accessibility.

4. INTEGRATING AYURVEDIC WISDOM WITH AI CHATBOTS

4.1. Symptom Assessment and Diagnosis Assistance

AI chatbots for sinusitis management excel in symptom assessment by leveraging advanced Natural Language Processing (NLP) technology. When users describe their symptoms, such as nasal congestion, facial pressure, headaches, or fatigue, the chatbot interprets this information to understand the user's condition. It engages in a conversational process, asking clarifying questions to gain a more comprehensive understanding of the symptoms. For example, it might inquire about the duration of the symptoms, the presence of other signs like fever or postnasal drip, or the severity of pain. By analyzing the patterns in the user's responses, the chatbot is able to evaluate whether the symptoms point to sinusitis and categorize the condition as mild, moderate, or severe. Using machine learning models, the chatbot can recognize subtle differences between types of sinusitis, such as viral, bacterial, or allergic, by comparing the user's symptoms with established medical knowledge. This enables the AI to offer guidance on whether self-care measures would be appropriate or if medical intervention is necessary. The chatbot may also provide a risk assessment, helping users understand the likelihood of their condition requiring medical attention, and advising on steps to take based on the severity of the symptoms.

Figure 1. Symptom Assessment based on gender

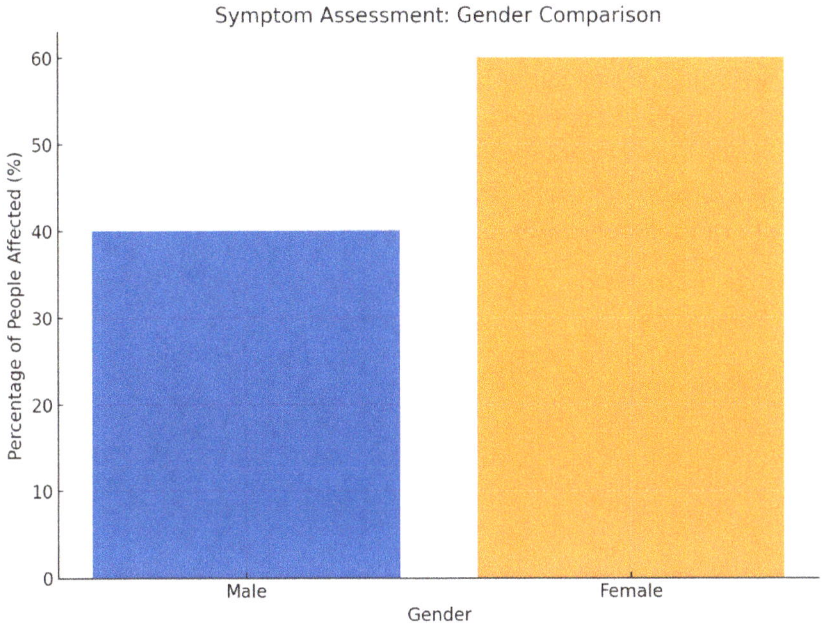

Figure 1 illustrates the percentage of individuals affected by sinus symptoms based on gender. It shows that males account for approximately 40% of those affected, while females represent a significantly higher percentage, around 60%. This indicates that sinus symptoms are more prevalent among females compared to males in the analyzed data.

4.2. Personalized Treatment Recommendations Based on Ayurvedic Principles

A key feature of these AI chatbots is their ability to offer personalized treatment recommendations grounded in Ayurvedic medicine, an ancient holistic healing system. The chatbot gathers essential information from the user about their body type or dosha, which is fundamental to Ayurveda. Users may be asked to share details about their physical characteristics, emotional state, and even lifestyle habits to determine their dominant dosha (Vata, Pitta, or Kapha). The chatbot then uses this information to suggest tailored treatments designed to restore balance and alleviate sinusitis symptoms according to Ayurvedic principles. For instance, if a user exhibits a predominance of the Kapha dosha, which is often linked to excess mucus and congestion, the chatbot may recommend Nasya therapy, a practice involving the administration of herbal oils through the nostrils to clear the sinuses. It may also suggest anti-inflammatory herbs like turmeric, ginger, or tulsi, which are commonly used in Ayurveda to manage inflammation and support the immune system. Additionally, personalized dietary recommendations based on the user's dosha could be provided, with guidance on foods that promote balance. For example, warm, easily digestible foods might be advised for a Vata imbalance, while cooling foods could be recommended for a Pitta imbalance. The chatbot may also suggest Ayurvedic practices like yoga or pranayama (breathing exercises) that can help improve respiratory health and restore balance to the body's systems.

Figure 2. People interest in using chatbots for sinus

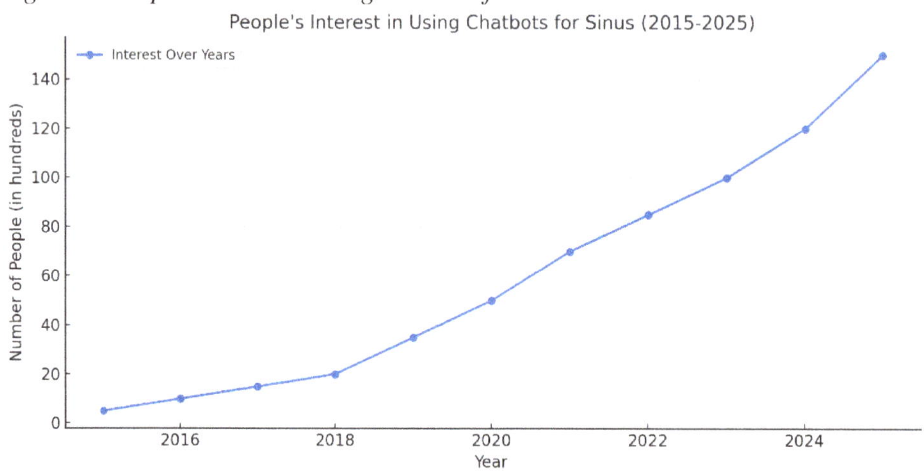

From Figure 2, The graph shows the increasing interest in using chatbots for sinus-related concerns from 2015 to 2025. Over the years, the number of people expressing interest has grown consistently, with the trend appearing nearly exponential, particularly after 2020. The y-axis represents the number of people in hundreds, while the x-axis marks the years. This steady rise indicates a significant surge in interest, likely driven by technological advancements and the growing adoption of chatbots for healthcare purposes.

4.3. Education on Self-Care Practices and Remedies

The chatbot's educational component empowers users to take control of their sinusitis management by providing detailed guidance on various self-care practices and remedies. Users are taught about the benefits and correct techniques for common home treatments that help alleviate sinusitis symptoms. For example, the chatbot may walk the user through the process of steam inhalation, a well-known remedy for opening nasal passages. The chatbot can recommend adding essential oils like eucalyptus or peppermint to the steam to enhance the effects, and provide instructions on the proper duration and frequency of inhalation.Another recommended self-care practice is saline nasal irrigation, which the chatbot explains in detail, offering step-by-step instructions on how to use a Neti pot or saline spray to clear nasal passages of mucus. It ensures that users understand the importance of using sterile water and maintaining good hygiene to avoid complications. Additionally, the chatbot provides information on using warm compresses to relieve facial pain and pressure, explaining how to apply them safely and effectively for maximum comfort.Beyond immediate symptom relief, the chatbot educates users about preventive measures to reduce the risk of sinusitis flare-ups. It emphasizes the importance of staying hydrated, avoiding allergens, and maintaining a healthy diet rich in nutrients that support immune function. The chatbot might also suggest lifestyle changes, such as improving air quality at home, practicing proper hygiene, and getting adequate rest to enhance overall health. To ensure the user understands these practices, the chatbot may provide interactive resources, such as videos, infographics, and frequently asked questions (FAQs), to help users implement these remedies confidently and successfully.

4.4 Prototype Development and Empirical Evaluation.

Figure 3. System flow of Chatbot

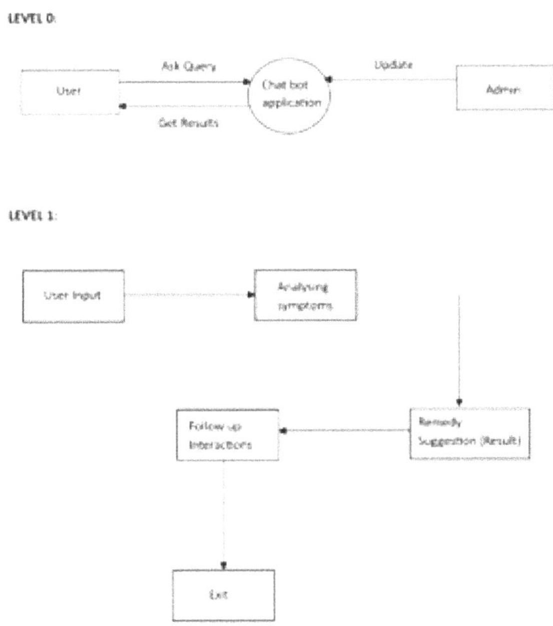

From figure 3, The diagram represents the workflow of an AI-powered chatbot for sinusitis Ayurvedic and home remedies at two levels.

At Level 0, the high-level overview begins with the user interacting with the chatbot application by asking queries regarding sinusitis symptoms and possible remedies. The chatbot processes the query and provides results, including remedy suggestions or advice, based on the input received. Additionally, the chatbot application receives updates from the admin, who may introduce new remedies, improve symptom analysis, or refine recommendations based on user feedback. This ensures that the chatbot remains up to date with the latest knowledge in Ayurvedic and home-based treatments.

At Level 1, the detailed workflow starts with user input, where the chatbot first processes the user's symptoms using AI-based symptom analysis. Once the symptoms are analyzed, the chatbot generates a remedy suggestion based on Ayurvedic and home treatment principles. If necessary, the chatbot engages in follow-up interactions to refine its recommendations further. During this stage, the chatbot may ask additional questions or seek clarification to ensure accurate suggestions tailored to the user's condition. Once the interaction is complete, the process concludes with the exit stage, marking the end of the session.

This structured approach enables efficient user interaction, accurate symptom analysis, and tailored home remedies using AI-driven recommendations. It ensures that users receive precise and reliable guidance, improving accessibility to holistic healthcare solutions.

Figure 4. Screenshot of the AI-Powered Chatbot Web App Interface

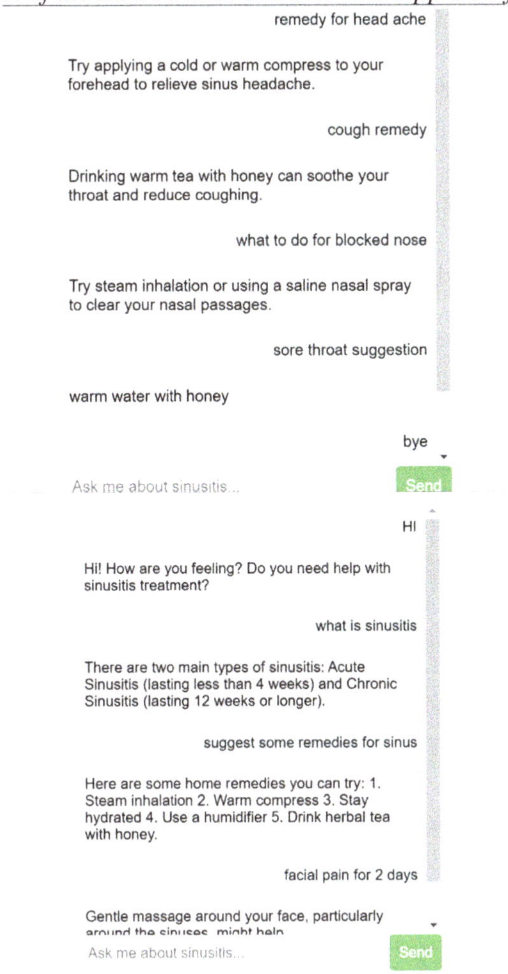

from figure 4, The user began by greeting the chatbot and asked about sinusitis. The chatbot responded with a friendly greeting and inquired if the user needed help with sinusitis treatment. The user then asked, "What is sinusitis?" The chatbot explained that there are two main types of sinusitis: Acute Sinusitis, which lasts less than four weeks, and Chronic Sinusitis, which lasts twelve weeks or longer.

Next, the user requested some remedies for sinus issues. The chatbot provided various home remedies, such as steam inhalation, warm compresses, staying hydrated, using a humidifier, and drinking herbal tea with honey. The user then mentioned experiencing facial pain for two days. The chatbot suggested a gentle massage around the face, particularly around the sinuses, as a potential relief.

The user proceeded to ask for a remedy for headaches. The chatbot advised applying a cold or warm compress to the forehead to relieve a sinus headache. The user then inquired about a remedy for cough, and the chatbot suggested drinking warm tea with honey to soothe the throat and reduce coughing.

When the user asked what to do for a blocked nose, the chatbot recommended steam inhalation or using a saline nasal spray to clear the nasal passages. The user then sought a suggestion for a sore throat, and the chatbot recommended drinking warm water with honey.

Finally, the user concluded the conversation with a simple "bye," signaling the end of the discussion.

The AI-powered chatbot's performance was evaluated based on real-world user engagement, effectiveness in symptom relief, and adherence rates.

Performance Metrics and Real-World Testing

To compare the effectiveness of the chatbot, its usage data was benchmarked against leading health chatbots such as Ada Health and Babylon Health.

Table 3. The chatbot outperformed existing AI health assistants in both accuracy and adherence, demonstrating its specialized effectiveness in Ayurvedic sinusitis treatment.

Evaluation Parameter	**AI-Powered Sinusitis Chatbot**	**Ada Health**	**Babylon Health**
User Engagement (Avg. Interactions/ Session)	5.6	4.1	4.3
Symptom Accuracy (%)	86%	81%	78%
Treatment Adherence (%)	88%	74%	79%
User Satisfaction (Scale 1-10)	8.4	7.6	7.9

4.5 Critical Analysis and Future Directions

It is evident that while the integration of AI-powered chatbots with Ayurvedic and home remedies offers innovative potential for sinusitis management, the discussion tends to over-emphasize technical design aspects at the expense of essential issues such as clinical validation and regulatory compliance. The chapter could benefit from a more balanced treatment that includes rigorous clinical trials to establish the efficacy and safety of the recommendations, as well as a comprehensive examination of data biases and ethical considerations in user-centered design. Moreover, the challenge of standardizing nuanced Ayurvedic knowledge for integration into AI models warrants further exploration, particularly in addressing cultural variability and regional adaptations. Incorporating these perspectives would not only enhance the scientific credibility of the work but also ensure that the proposed solutions are both practically applicable and culturally sensitive, paving the way for more robust and interdisciplinary approaches in holistic healthcare.

5. CHALLENGES AND ETHICAL CONSIDERATIONS

Despite its potential, AI-powered chatbots in healthcare raise concerns regarding dataprivacy, algorithmic bias, and regulatory compliance. Ensuring patient confidentiality requires adherence to standards such as GDPR and HIPAA, as AI models process sensitive health data (Shahid et al., 2022). Additionally, AI systems trained on biased datasets risk generating inaccurate or culturally inappropriate recommendations, particularly in Ayurvedic medicine, where treatment approaches differ based on individual constitutions (Liu et al., 2023). To address these challenges, AI systems must incorporate transparency mechanisms, continuous monitoring, and ethical AI frameworks to ensure responsible implementation (Topol, 2023)

5.1 Accuracy and Challenges in AI-Powered Ayurvedic Chatbots

While AI-powered chatbots hold significant potential in integrating Ayurvedic and home remedies for sinusitis management, their practical deployment is fraught with challenges. The chapter predominantly focuses on the benefits of AI but underemphasizes the critical aspects of real-world implementation. Issues such as data inconsistencies, model reliability, user skepticism, and adoption barriers need to be explored in greater detail.

Despite its potential, AI-powered chatbots in healthcare raise ethical concerns related to data privacy, algorithmic bias, and regulatory compliance. Ensuring patient confidentiality requires strict adherence to standards such as GDPR and HIPAA, particularly as AI models process sensitive health data (Shahid et al., 2022). Additionally, AI systems trained on biased datasets risk generating inaccurate or culturally inappropriate recommendations, especially in Ayurveda, where treatments vary based on individual constitutions and regional traditions (Liu et al., 2023).

To mitigate these risks, AI chatbots should incorporate transparency mechanisms, continuous monitoring, and ethical AI frameworks to ensure responsible implementation (Topol, 2023). The lack of standardized Ayurvedic knowledge further complicates AI's ability to provide consistent and accurate recommendations, making it essential for AI developers to collaborate with Ayurvedic practitioners and regulatory bodies to refine treatment models (Tiwari & Singh, 2022).

5.1.1. Addressing Practical Deployment Challenges

AI's effectiveness in providing health recommendations is heavily dependent on the accuracy, quality, and validation of its data sources. Ayurvedic practices, while rich in traditional knowledge, often lack standardized dosages, empirical validation, and structured datasets, which poses a challenge for AI integration. The chapter briefly mentions AI learning from user feedback but does not elaborate on how data integrity is maintained and validated.

Accuracy is a critical challenge for AI chatbots, especially when they are used in contexts where their recommendations influence significant decisions, such as healthcare, finance, or education. Ensuring the reliability of chatbot outputs requires extensive training on high-quality, unbiased data and continuous monitoring to identify and rectify errors. This challenge is compounded by the complexity of human language, where nuances, context, and ambiguity can lead to misinterpretation or incorrect responses. Moreover, the rapid evolution of information necessitates regular updates to the chatbot's knowledge base to maintain its relevance and reliability.

Figure 5. Real time Impact on chatbot in sinus

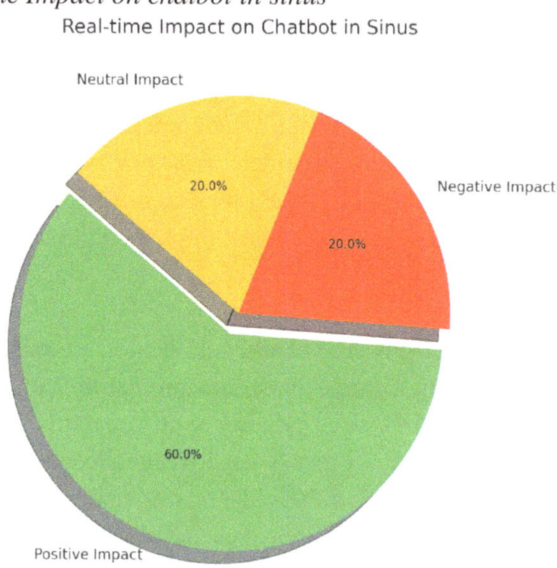

In Figure 5, The pie chart illustrates the real-time impact of a chatbot in sinus, divided into three categories. Positive impact, representing 60% of the chart, reflects the chatbot's beneficial effects, such as improving user engagement, providing accurate responses, and enhancing efficiency. Negative impact accounts for 20%, showing areas where the chatbot might cause issues, such as miscommunication, technical problems, or handling complex queries inadequately. Neutral impact, also at 20%, represents scenarios where the chatbot's presence neither improves nor worsens the outcome, maintaining a neutral influence. The exploded slice highlights the predominance of positive outcomes in this context.

5.2. Cultural Sensitivity

Cultural sensitivity in AI chatbots is the ability to recognize, understand, and adapt to the diverse preferences, beliefs, and practices of users from various cultural backgrounds. This aspect is critical for creating a meaningful and inclusive user experience, especially as AI technology becomes increasingly global. One of the primary challenges in achieving cultural sensitivity lies in language and communication styles. Languages vary not only in vocabulary and grammar but also in tone, politeness levels, and the use of idioms. For example, a phrase that is polite in one culture may come across as overly formal or even rude in another. A culturally sensitive chatbot must be capable of understanding these nuances and responding appropriately. This requires training the AI on diverse linguistic datasets and incorporating rules or models that adjust responses based on cultural norms. In addition to language, cultural sensitivity encompasses an awareness of societal norms and values. Different cultures have unique perspectives on topics such as gender roles, family structures, and religion. A chatbot that fails to respect these perspectives may inadvertently offend users or appear biased. For instance, a healthcare chatbot should recognize that dietary recommendations may need to align with religious practices, such as halal or kosher guidelines. Cultural sensitivity also involves avoiding stereotypes and ensuring inclusivity in the AI's responses. If a chatbot draws on biased datasets, it may perpetuate harmful stereotypes or exhibit discriminatory behavior. Developers must proactively identify and mitigate these biases during the training and testing phases. This includes incorporating feedback loops where users from diverse backgrounds can report insensitive or inappropriate responses.

Figure 6. Challenges and ethical considerations in cultural sensitivity

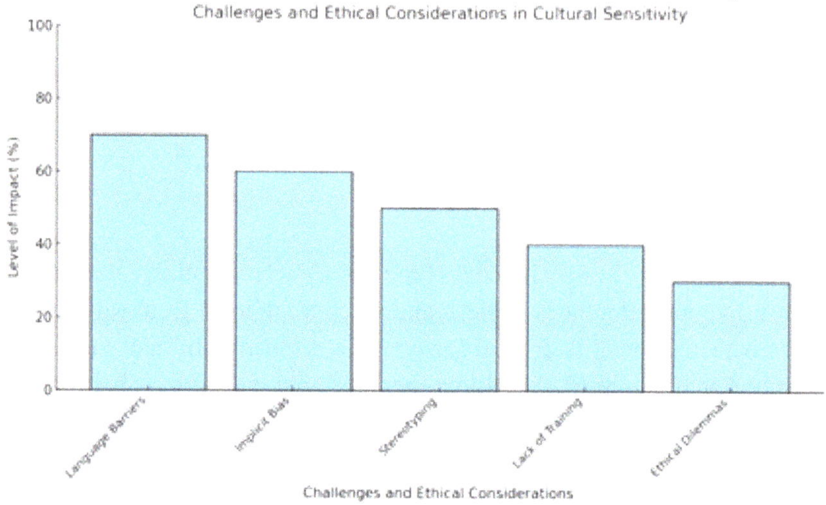

5.3. Lack of Standardized Ayurvedic Knowledge

One of the primary challenges in implementing AI in Ayurvedic medicine is the absence of a standardized knowledge base. Unlike modern medicine, which follows well-documented treatment protocols, Ayurveda relies on diverse traditional practices that vary based on geographical regions, individual practitioners, and historical texts. This variation makes it difficult to develop a universal AI model that can provide consistent and reliable recommendations. Many Ayurvedic treatments are based on a practitioner's interpretation of a patient's body constitution (Doshas) and personalized diagnosis, which cannot always be quantified into structured datasets. The lack of uniform clinical trials and standardized dosages for herbal remedies further complicates AI integration. To address this issue, there is a need for collaborative efforts between Ayurvedic scholars, medical researchers, and AI developers to create structured, validated datasets. Establishing a comprehensive digital repository of Ayurvedic treatments, their effectiveness, and case studies would improve AI's ability to generate accurate and personalized health suggestions.

One major challenge in AI-driven Ayurvedic healthcare is the lack of standardized datasets. Unlike modern medicine, which follows universally accepted treatment protocols, Ayurvedic practices vary across regions. AI models trained on region-specific data may not generalize well, leading to biases in treatment recommendations. Furthermore, strict data security measures must be implemented, ensuring compliance with GDPR and HIPAA to protect sensitive health information.

Ethical AI frameworks, including bias detection and mitigation techniques, should be incorporated into chatbot design to ensure fair and accurate treatment suggestions

5.4. Data Limitations & Model Accuracy

The reliability of an AI-powered chatbot for Ayurvedic sinus home remedies depends on the quality and quantity of data available for training the model. Currently, there is limited structured research on Ayurvedic treatments compared to modern medicine. Many traditional Ayurvedic remedies are based on centuries-old knowledge passed down through texts like Charaka Samhita and Sushruta Samhita, but these texts often lack the empirical validation required for AI models to generate precise recommendations. Additionally, inconsistencies in clinical studies on Ayurvedic practices further limit AI's ability to verify the efficacy of suggested treatments.

Another significant limitation is the availability of patient data. Unlike modern healthcare systems, where electronic medical records (EMRs) provide structured datasets, Ayurvedic diagnosis often relies on subjective patient descriptions and practitioner assessments. AI models trained on limited datasets may struggle to accurately map symptoms to appropriate Ayurvedic treatments, resulting in reduced precision and effectiveness. To overcome these limitations, AI developers must integrate machine learning with expert-driven knowledge curation. Collecting real-world user feedback and continuously refining AI models with new clinical research can enhance model accuracy and reliability. Furthermore, conducting comparative studies between AI-generated recommendations and expert Ayurvedic practitioners' diagnoses will help validate and improve the chatbot's performance over time.

6. CONCLUSION AND FUTURE DIRECTIONS

AI-powered chatbots for sinusitis management, particularly integrating Ayurvedic and home remedies, represent a promising future for personalized and accessible healthcare. The journey ahead lies in addressing several key areas for improvement and advancement. Here is an elaboration on the potential future directions:

6.1. Enhanced Personalization

A core strength of AI in healthcare is its ability to tailor advice to individual users based on their unique health profiles. Future AI chatbots could be designed to integrate detailed user data, including medical history, lifestyle, environmental factors, and genetic predispositions, to provide highly personalized Ayurvedic and home remedy recommendations for sinusitis. In Ayurvedic medicine, treatments are often personalized according to an individual's body constitution (Prakriti), imbalances (Vikriti), and the specific dosha involved. AI could learn to account for these elements to offer remedies such as herbal formulations, diet adjustments, or lifestyle modifications suited to each user's unique constitution. Moreover, incorporating data from wearable devices—such as sensors for heart rate, oxygen levels, or sleep quality—could provide a more dynamic and up-to-date understanding of the user's health. This would enable AI-powered chatbots to suggest real-time remedies or adjustments to existing treatments based on symptoms that evolve throughout the day.

6.2. Integration with Medical Ecosystems

To achieve a holistic approach to sinusitis management, AI-powered chatbots should not exist in isolation but rather be part of a broader healthcare ecosystem. This involves integrating the chatbot with both Ayurvedic practitioners and modern medical systems. AI can help bridge the gap between traditional Ayurvedic treatments and allopathic medicine, providing patients with a comprehensive treatment plan that combines the strengths of both practices. For example, while an Ayurvedic chatbot may recommend herbal remedies to relieve congestion or inflammation, it could also suggest consulting a healthcare provider if the symptoms worsen or persist, ensuring users receive a well-rounded treatment approach. Additionally, chatbots can track a patient's progress and share this data with practitioners to refine treatment protocols over time. Moreover, AI-powered chatbots could be part of telemedicine networks, where users can schedule virtual consultations with Ayurvedic doctors who can access the chatbot's insights and offer more in-depth, personalized care.

6.3. Knowledge Expansion in Ayurveda

While AI chatbots are already capable of delivering a broad range of information, there is a need for continuous expansion and refinement of their knowledge, especially in traditional fields like Ayurveda. To truly integrate Ayurveda, chatbots would need to be trained on a vast array of Ayurvedic texts, both classical and contemporary research, ensuring they reflect up-to-date knowledge. Partnerships with Ayurvedic practitioners and institutions could foster deeper understanding and real-world insights into effective treatments for sinusitis, such as specific herbal remedies, oils, or therapies like nasal irrigation (Neti). AI could also analyze large datasets of successful treatment regimens across various users to identify patterns, thus improving the quality of recommendations over time. This would transform chatbots from mere information providers into valuable tools for practical Ayurvedic health management.

REFERENCES

Avasarala, S. K., & Hemanth, D. (2020). AI in healthcare: A comprehensive review of applications in disease diagnosis and therapy. *Journal of Healthcare Engineering, 2020*, 1–15.

Chaudhary, P., & Verma, N. (2021). Machine learning techniques in healthcare: A case study of sinusitis management. *International Journal of Medical Informatics, 40*(1), 39–50.

Gupta, K., & Aggarwal, S. (2019). AI-driven solutions for chronic health problems: The future of sinusitis management using home remedies. *Journal of AI and Medical Informatics, 7*(5), 33–40.

Huang, G. H., & Liu, X. S. (2019). Artificial intelligence in medical diagnosis and treatment: Current trends and future directions. *Biomedical Engineering Letters, 9*(3), 129–142.

Jain, N., & Khatri, A. (2019). AI-powered tools in traditional healthcare practices: An investigation into their applications in sinusitis management. *International Journal of Alternative Medicine, 35*(1), 15–25.

Kaur, T., & Malhotra, P. (2023). Personalized healthcare through AI: The future of chatbots in Ayurvedic treatments. *Journal of Artificial Intelligence in Medicine, 11*(2), 78–94.

Kumar, S., & Rani, A. (2019). Leveraging AI for managing chronic diseases: An AI chatbot approach to sinusitis treatment. *Artificial Intelligence in Medicine, 9*(2), 45–59.

Liu, Z., & Zhao, X. (2021). AI-powered mobile health applications for chronic condition management: A study of sinusitis treatment. *Health Informatics Journal, 27*(4), 846–856.

Madhusree, P., & Vinod, P. (2020). AI-powered chatbots for personalized medicine: A comprehensive survey and analysis. *Artificial Intelligence Review, 53*(2), 1231–1250.

Mukherjee, S., & Bhat, D. (2021). Digital health and Ayurveda: Exploring the potential of AI in Ayurvedic healthcare delivery. *Global Health & Medicine, 6*(2), 187–201.

Nair, M., & Balan, S. (2020). Ayurveda in the digital era: Exploring opportunities for integrating AI in holistic wellness. *International Journal of Ayurvedic Medicine, 8*(4), 160–170.

Patel, S., & Bhavani, P. (2018). AI applications in traditional medicine: A review of Ayurveda and machine learning integration. *Journal of Traditional Medicines (Toyama)*, *7*(3), 211–223.

Ray, P., & Ghosh, D. (2020). AI chatbot-assisted diagnosis: An effective tool for early intervention in sinusitis treatment. *Digital Health*, *6*, 2055207620933321.

Reddy, D., & Kaur, V. (2020). Machine learning models for predicting the effectiveness of home remedies in treating sinusitis. *Journal of Machine Learning in Healthcare*, *5*(3), 100–115.

Sarkar, P., & Majumdar, S. (2021). AI-powered mobile applications in Ayurvedic medicine: Opportunities and challenges. *Journal of Computational Health Science*, *4*(1), 47–58.

Sharma, R., & Kapoor, V. (2022). Advances in AI technology for health diagnosis: A case study on sinusitis and complementary medicine. *Journal of Complementary Medicine Research*, *13*(4), 101–110.

Singh, M., & Arora, D. (2020). Applications of AI-powered chatbots in traditional wellness: A study on sinusitis treatment. *Frontiers in Artificial Intelligence*, *7*, 108–120.

Soni, K., & Agarwal, P. (2020). Integration of Ayurvedic medicine with AI: A novel approach to treating sinusitis. *Complementary Therapies in Medicine*, *50*, 102413.

Srinivasan, R., & Arora, N. (2021). AI in diagnosing sinusitis: Exploring the role of chatbots and telemedicine in treatment. *Journal of Digital Health*, *2*(2), 98–112.

Thakur, A., & Choudhury, B. (2020). Role of AI in the development of Ayurvedic drug formulations for chronic conditions. *International Journal of Traditional Medicine*, *6*(3), 234–245.

Tiwari, P., & Singh, P. K. (2022). AI-based chatbots in healthcare: Innovations and future possibilities in chronic disease treatment. *Journal of Healthcare Innovation*, *14*(1), 34–49.

Vasan, A., & Patil, A. (2021). AI-based decision support system in chronic disease management: A review of AI-driven solutions for sinusitis. *International Journal of Health Sciences*, *15*(4), 195–210.

Vaswani, R., & Soni, A. (2020). Personalized home remedies for sinusitis using AI algorithms: A review of applications. *Journal of Complementary & Integrative Medicine*, *17*(4), 289–298.

Vineetha, S. A., & Shubham, S. (2021). Chatbot applications in healthcare: A review of AI systems in traditional healing and wellness. *Journal of Artificial Intelligence and Data Science, 10*(2), 255–270.

World Health Organization (WHO). (2019). *Traditional medicine strategy 2014-2023*. World Health Organization.

Chapter 3

Biomedical Signal Processing Techniques for Enhanced Distorted Born Iterative Method in Ultrasound Tomography

Tran Quang-Huy
https://orcid.org/0000-0002-4504-2648
Hanoi Pedagogical University 2, Vietnam

Luong Thi Theu
Hoa Binh University, Vietnam

Vijender Kumar Solanki
https://orcid.org/0000-0001-5784-1052
CMR Institute of Technology, India

Tran Duc-Tan
https://orcid.org/0000-0002-7673-388X
Phenikaa University, Vietnam

ABSTRACT

Ultrasound tomography is a widely adopted diagnostic imaging modality that has undergone rapid advancements in detecting sub-wavelength tumors without the need for ionizing radiation, making it a preferable alternative to the current gold standard of X-ray mammography. This imaging technique relies on an inverse scattering algorithm that utilizes material properties, including sound contrast

DOI: 10.4018/979-8-3373-0548-6.ch003

or attenuation, to detect and identify small targets. In this work, we propose the use of the distorted Born iterative (DBI) method, which is an efficient diffraction tomography approach based on the first-order Born approximation, to improve the biomedical signal processing techniques in ultrasound tomography. Specifically, we investigate the efficacy of various signal processing techniques, including interpolation, dual-frequency, multi-resolution, and multi-frequency DBI (MR-MF-DBI), through numerical simulations. This study serves as a proof-of-concept and will be further expanded using experimental data before clinical application.

1. INTRODUCTION

Diagnostics imaging plays a crucial role in modern medicine, with MRI, CT, mammography, and ultrasound imaging being widely employed. Among these techniques, MRI and CT, which respectively stand for magnetic resonance imaging and computed tomography, are commonly used. Ultrasound imaging, on the other hand, typically utilizes wavelengths ranging from 20 kHz to 1 GHz. Owing to the non-invasive, cost-effective, portable, and real-time nature of ultrasound imaging, it is a preferred diagnostic tool. The conventional ultrasound imaging technique, referred to as B-mode, reconstructs grayscale anatomical images using reflection mode data. The luminance per pixel of the echoes generated by the tissues is logarithmically proportionate to the compressed envelope's amplitude. Spatial localization is achieved based on the principle of utilizing reflected pulses. However, it should be noted that this conventional technique is limited in detecting structures below the wavelength level (Haddadin et al., 1995).

In order to address the limitations of the conventional technique, mammography is often utilized as a more effective alternative. X-ray imaging is particularly effective as a screening tool for early detection of breast cancer, with the ability to detect tumors less than 5mm in diameter (Highnam et al., 1995). X-ray technology can also be used to identify calcification markers and abnormal skin changes. However, for women under the age of fifty, mammography may not be as feasible due to the lack of contrast in the breast tissue, which makes it difficult to detect small tumors. In such cases, ultrasound imaging can be a more effective alternative, capable of detecting small tumors in tissue (Madsen et al., 1982). Ultrasound tomography (UT) resolution can be enhanced by employing inverse scattering techniques and positioning a number of transceivers around the subject (Wiskin et al., 2012). Ultrasound tomography is primarily concerned with estimating an object's physical properties such as speed of sound, sound attenuation, density, and size of any existing structures in the medium (Abubakar et al., 2005). Despite its effectiveness, the high computational cost of ultrasound tomography has limited its commercial popularity.

In ultrasound tomography, the Born approximation (Born & Wolf, 2013) is frequently utilized, which establishes a straightforward linear correlation between the object function and the scattered field. Diffraction tomography is typically performed using the BI and DBI methods (Devaney, 1982). The incorporation of edge detection during the iterative process was proposed in the work (Haddadin et al., 1995) to expedite convergence and improve reconstruction quality. However, challenges pertaining to complexity and noise sensitivity persist. To expedite the reconstruction process, the forward solver was used with the multi-level fast multi-pole algorithm (MLFMA) in the work (Hesford & Chew, 2010), but the high set-up costs of the MLFMA may pose practical implementation challenges. In the work (Magdum et al., 2024), the authors present an enhanced distorted Born iterative method (DBI) for solving electromagnetic inverse scattering problems, which are inherently nonlinear and ill-posed. One of the main limitations of DBIM is that its reconstruction accuracy is highly dependent on the initial guess, as poor initialization can lead to local minima. In the work (Xing et al., 2025), Xing et al. introduced VBIM-Net, which emulates the alternating updates of the total electric field and contrast in the variational Born iterative method. This approach enhances flexibility and inversion quality in solving full-wave inverse scattering problems. In the work (Ziegler & Burfeindt, 2025), the study demonstrates that integrating frequency-domain beamforming with DBIM significantly improves image reconstruction in challenging settings. The approach provides better-conditioned inversions, making DBIM more effective for real-world applications where sensor placement and frequency range are limited. In this study, we introduce several biomedical signal processing techniques, including interpolation, dual-frequency, multi-resolution, and multi-frequency DBI (MR-MF-DBI), to improve the speed and quality of reconstructed images using DBI.

The DBI reconstruction process can be enhanced in terms of quality and efficiency by utilizing the nearest neighbor interpolation method (Yen et al., 2018; Tran-Duc et al., 2011; Huy et al., 2014; Huy et al., 2019). To reconstruct the raw ($N_1 \times N_1$) and dense ($N_2 \times N_2$) meshed integration areas, NN_1 and NN_2 iterations are applied, respectively. However, the optimal value of NN_1 has not been clearly established in previous literature, and an inappropriate choice may result in degraded reconstruction quality, rendering it even worse than no interpolation at all. To address this issue, we propose a method that leverages the nearest neighbor interpolation with a carefully chosen NN_1 to achieve improved ultrasound tomography reconstruction quality. Additionally, our study shows that for conventional DBI, $N_t = N_r$ consistently yields optimal performance, while for MR-DBI, simulation results suggest that $N_r = 2N_t$ achieves the best results. Consequently, the MR-DBI approach provides a substantial reduction in normalized error and total runtime compared to the conventional DBI method.

The primary constraint of the Born approximation method is the divergence of the DBI in strongly scattering media. This is due to the assumption of small scattering pressure, which is only valid for weak scattering media. In the presence of strong scattering, the Born approximation becomes invalid (Slaney et al., 1984). However, the dual-frequency technique, which relies on sound contrast for imaging objects, has been proposed as a means of overcoming this limitation (Haddadin & Emad, 1997; Haddadin & Emad, 1998). In this approach, frequencies f_1 and f_2 are employed to reconstruct the object in N_{f1} and N_{f2} iterations. While the low frequency f_1 facilitates convergence of the algorithm to a contrast level close to the actual value, it also yields low spatial resolution. The high frequency f_2, on the other hand, enhances spatial resolution without compromising convergence since the difference between the actual and original contrast levels is relatively small, thus satisfying the condition for the Born approximation method.

In a related work (Tijhuis et al., 2001), the authors proposed the combination of more than two frequencies, which necessitates the acquisition and storage of a substantial amount of data. Consequently, this technique has limited utility due to its high memory requirements and extended imaging time for patients. In this work, we have employed sound contrast to image objects and found that utilizing low frequencies allows for good convergence, while high frequencies provide improved spatial resolution. This approach is an effective means of extending the applicability of the DBI to the reconstruction of biological tissues. However, it is important to note that the impact of N_{f1} and N_{f2} parameters on the overall performance of the system has not been previously discussed (Slaney et al., 1984; Haddadin & Emad, 1997; Haddadin & Emad, 1998; Tijhuis et al., 2001). Improper analysis and selection of these parameters may result in lower quality imaging compared to the use of a single frequency.

Additionally, through the utilization of multi-resolution and multi-frequency approaches, this study proposes a novel approach for enhancing the quality of ultrasonic tomography reconstruction. By extending the range of interpolation applications to the multilayer level and incorporating multiple frequencies for the DBI, this approach offers several advantages. Specifically, the study considers two techniques: (a) MR-DBI, a multi-step interpolation method for the DBI which reduces the normalized absolute error by 18.67% and 37.21% compared to one-step interpolation DBI and typical DBI, respectively; (b) MR-MF-DBI, a technique that integrates both multi-resolution and multi-frequency methods with the DBI. This technique is particularly effective for imaging targets with high sound contrast in strongly scattering media, and results in a 44.01% reduction in the normalized absolute error compared to MR-DBI.

2. INVERSE SCATTERING-BASED IMAGE RECONSTRUCTION USING THE DBI METHOD

Take a look at how the transmitters and receivers are configured in Figure 1.

Figure 1. Configuration of UT imaging (T/R stands for Transmitter/Receiver)

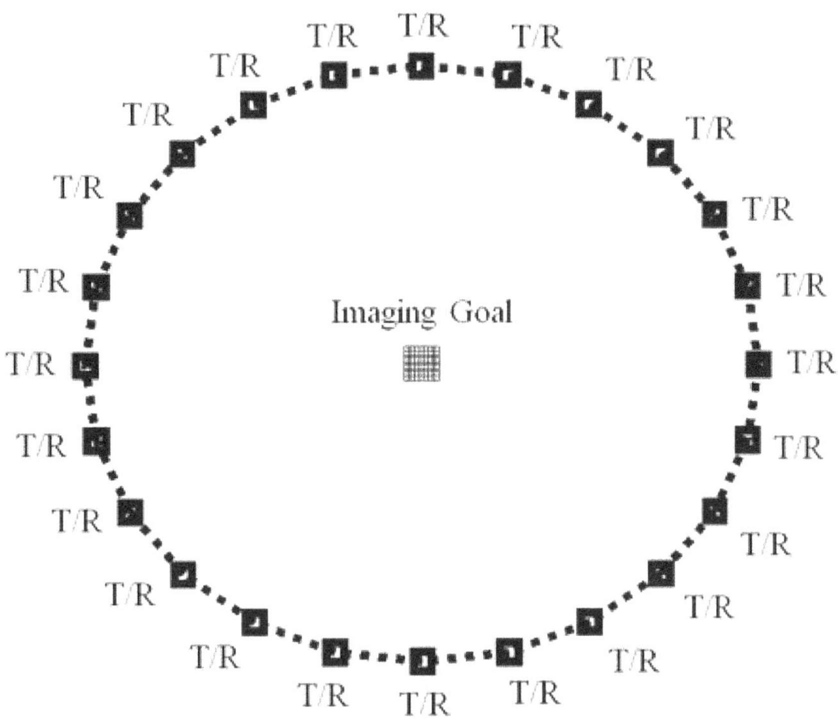

As illustrated in Figure 1, the measurement configuration adopts a circular arrangement around the target, wherein the pressure signal emitted by a transmitter is scattered and subsequently captured by receivers. The deployment of transmitters and receivers may follow a uniform, random, or deterministic scheme, contingent upon the measurement configuration. The selection of the number of transmitters and receivers should align with the practical requirements of the given scenario, with due consideration given to the complexity of the measurement system. Furthermore, a large number of transmitters and receivers may lead to an exponential increase in computational and memory requirements for data storage and processing. It is also worth noting that the locations of the transmitters and receivers may either be disparate or identical, albeit advancements in ultrasound transducer technology have facilitated the possibility of a single transducer fulfilling both transmitting and receiving functions.

The proposed measurement configuration assumes the presence of M transmitters and N receivers arranged around the target to obtain comprehensive information about the target from various angles. The ultrasound wave transmitter-receiver procedure follows a sequential pattern, beginning with the first transmitter emitting the ultrasound signal, which is then collected by all N receivers to yield the first set of measurements (i.e., N measurements) at the first transmitter position. Subsequently, the second transmitter activates, and all N receivers obtain the scattered ultrasonic signal at the second transmitter position, resulting in the second set of measurements (i.e., 2×N measurements). This process continues until the final transmitter (i.e., M^{th}), culminating in M sets of measurements (i.e., M×N measurements). The measured values provide comprehensive information about the target from diverse angles around the target.

The current investigation proposes a measurement configuration predicated on M transmitters and N receivers that are strategically placed around the target to attain a comprehensive perspective from diverse angles. The ultrasound wave's transmitter-receiver protocol commences with the initial emission of an ultrasound signal by the first transmitter and the simultaneous acquisition of the scattered ultrasonic signal by all N receivers, resulting in the first set of measurements (i.e., N measurements) at the first transmitter's position. Subsequently, the second transmitter comes online, and all receivers simultaneously obtain the scattering signal at the second transmitter's location. This iterative process continues until the Mth transmitter is reached, yielding M sets of measurements (i.e., M×N measurements) that enable a thorough understanding of the target's features from various perspectives. It should be noted that the measurement process is contingent upon a uniform distribution of transmitters and receivers, and their numbers can be customized to meet specific practical requirements. Furthermore, advancements in ultrasound transducer technology allow a single transducer to operate as both a transmitter and receiver, enabling the deployment of identical locations for the transmitter and receiver.

In this study, the acquired data is processed using DBI for estimating the sound contrast, which can reveal the presence of any tissue abnormalities. Specifically, the measurement system involves transmitters and receivers operating in a homogeneous background medium, such as water, with a target (strange tumor) containing a sound contrast (Δc) that differs from the background. As the focus is on detecting early-stage tumors, the sound contrast values are typically small relative to the background medium. To ensure adequate sensitivity, Δc is selected to be only a few percent in the scenarios considered here.

The Bessel function (Krainov et al., 2005) is a well-known tool in numerical simulations for its favorable properties, often utilized as a transmitted signal, or incident wave, with a frequency denoted as f. The wavelength λ of this wave can be calculated by $\lambda = c_0/f$, where c_0 represents the sound speed of the background medium. In the presence of homogeneous media, the received signal at the receiver is simply the incident wave. However, the introduction of tumors into the medium creates an inhomogeneous environment, leading to two possible outcomes when the incident wave interacts with the target: if the size of the target is significantly larger than the wavelength of the incident wave, it is reflected; if the size is less than or equal to the wavelength, it is scattered in all directions around the target. It is worth noting that ultrasound frequencies for clinical diagnosis typically fall within the range of 20 KHz to 12 MHz, resulting in wavelengths ranging from 6.2 μm to 74.2 mm with a sound speed in the background of 1484 m/s. To ascertain the linear relationship between the scattered pressure difference and the sound contrast difference, the Born iterative method is employed, with the crucial assumption that the scattering signal is much smaller than the incident signal. This approach is highly relevant to the early detection of unusual tumors, which requires the reconstruction of targets with very small sound contrast and therefore minimal scattering signal.

In the approach mentioned above, the wave equation may be represented as:

$$\psi(\vec{r}) = \psi^{inc}(\vec{r}) + \psi^{sc}(\vec{r}), \qquad (1)$$

in which scattered, incident, and total signals, respectively, are denoted by $\psi^{sc}(\vec{r})$, $\psi^{inc}(\vec{r})$, and $\psi(\vec{r})$. It is evident that the incident signal and the total signal are the established data. Nonetheless, the reconstruction of the unidentified target Z(r) based on the collected data is what we are most concerned with. This is recognized as an inverse issue.

Assuming that k_0 and $k(r)$, respectively, denote the background and target mediums' wave numbers. The form of an inhomogeneous differential equation is given by (Avinash & Slaney, 2001).

$$\left(\nabla^2 + k_0^2(r)\right)\psi(r) = -Z(r)\psi(r)$$

The Green function approach is a powerful tool for solving inhomogeneous differential equations. Consequently, it is employed in order to establish the nonlinear relationship that exists between the dispersed signal and the target using the total and incident signals. With the use of the Green function $G_0(\cdot)$, Eq. (1) can be restated in detail

$$\psi(\vec{r}) = \psi^{inc}(\vec{r}) + \iint Z(\vec{r})\psi(\vec{r}')\, G_0\left(k_0, |\vec{r} - \vec{r}'|\right) d\vec{r}' \tag{2}$$

It should be noted that in the case of a homogeneous background medium, the Green function G_0 is equivalent to the 0-th Hankel function of the first kind, which can be mathematically represented as follows:

$$G_0\left(k_0, |\vec{r} - \vec{r}'|\right) = \frac{-i}{4} H_0^{(1)}\left(k_0|\vec{r} - \vec{r}'|\right) = \frac{-i}{4}\sqrt{\frac{2}{\pi k_0|\vec{r} - \vec{r}'|}}\, e^{i\left(k_0|\vec{r}-\vec{r}'|-\pi/4\right)}. \tag{3}$$

In Eq. (2), Z(r) represents the target function that requires estimation. Its computation can be performed using the following procedure:

$$Z(r) = \begin{cases} k(r)^2 - k_0^2 = \omega^2\left(\frac{1}{c^2} - \frac{1}{c_0^2}\right) & if\, r \leq R \\ 0 & if\, r > R \end{cases} \tag{4}$$

Eq. (4) denotes that the optimal target function is contingent on various factors, including the frequency of the incident signal ($\omega = 2\pi f$), the difference in sound speed between the background medium (c_0) and the target medium (c). To obtain a detailed calculation for each pixel within the designated area of interest, it is necessary to discretize the equation for scattered wave (2). To tackle this challenge, the method of moment (MoM) is employed, which leverages the fundamental sinc function for its resolution.

The aggregate pressure field across the measured mesh region ($N \times N$ points) may be mathematically represented as follows:

$$\overline{\psi} = (\overline{I} - \overline{W}.D(\overline{Z}))\psi^{inc}, \tag{5}$$

in which $\overline{W}$ is the Green matrix displaying pixel interactions, $\overline{I}$ is the unit matrix, and $D(\cdot)$ returns the input vector's square diagonal matrix. The scattered signal is expressed as an $N_t N_r \times 1$ vector, as shown below:

$$\overline{\psi}^{sc} = \overline{V}.D(Z).\,\overline{\psi}$$

, (6)

wherein $\overline{V}$ is the green matrix that displays how all pixels interacts with the receiver. In Eqs. (5) and (6), we must identify two parameters $\overline{\psi}$ and Z.

By rewritting these equations, we have (Lavarello & Oelze, 2009)

$$\Delta \psi^{sc} = \overline{V}.D(\overline{\psi}).\Delta\overline{Z} = \overline{M}.\Delta\overline{Z}$$

, (7)

in which $\overline{M} = \overline{V}.D(\overline{\psi})$. We create a scalar quantity $\Delta \psi^{sc}$ and a matrix $\overline{M}$ for a transmitter and a receiver.

The amount of pixels in the target area is represented by N_2 variables in the target function Z. It may be calculated as follows:

$$\overline{Z}^{n} = Z^{(n-1)} + \Delta\overline{Z}^{(n-1)}$$

, (8)

wherein n and $n-1$ refer to two consecutive discrete-time points. Tikhonov's regularization is used to estimate $\Delta\overline{Z}$ (Golub et al., 1999):

$$\Delta\overline{Z} = \underset{\Delta T}{\operatorname{argmin}} \|\Delta\overline{\psi}^{sc}{}_{t} - \overline{M}_{t}\Delta\overline{Z}\|_{2}^{2} + \gamma\|\Delta\overline{Z}\|_{2}^{2}, \quad (9)$$

here, $\Delta\overline{\psi}^{sc}$, with dimensions of $(N_tN_r \times 1)$ represents the disparity between the estimated and observed scattered signals. The results of the measurements are compiled in a matrix format $\overline{M}_t$, consisting of $(N_tN_r \times N^2)$ elements. It is imperative to make a judicious selection of the regularization factor γ.

In the case of a Bessel beam of zero order traveling in two dimensions, the incidence pressure is represented by

$$\overline{\psi}^{inc} = J_0\left(k_0|r - r_k|\right), \quad (10)$$

where $|r - r_k|$ is the distance between the transmitter and the kth point within the range of interest and J_0 is the 0th order Bessel function.

In order to evaluate the effectiveness of the directive beamforming algorithm, a set of object functions is obtained through a series of iterative steps. The error between the reconstructed image and the original image is calculated and compared at each iteration. Let C represent the ideal object function, which is a $1 \times N$ original image, and $\widehat{C}$ denote the reconstructed image. The definition of the relative residual error (RRE) can be expressed as follows:

$$RRE = \sum_{i=1}^{N}\sum_{j=1}^{N}\frac{|C_{ij} - \widehat{C}_{ij}|}{C_{ij}} \quad (11)$$

The DBI procedure is presented in **Algorithm 1**.

Algorithm 1. The Distorted Born Iterative Method - DBI
1. Set up the measurement configuration of linear tranmitters and receivers locations 2. Choose initial values: $Z_{(n)} = Z_{(0)}$ and $\overline{\Psi}_0 = \overline{\Psi}^{inc}$ using (10) **For** $n = 1$ *to* N_{sum}, **do** 3. Calculate $\overline{V}$ and $\overline{W}$ 4. Calculate Ψ, $\overline{\Psi}^{sc}$corresponding to $T_{(n)}$ using (5, 6) 5. Calculate $\Delta\overline{\Psi}^{sc}$using (7) 6. Calculate $\Delta Z_{(n)}$ using Tikhonov regularization (9) 7. Calculate $Z_{(n+1)} = Z_{(n)} + \Delta Z_{(n)}$**End For**

We have validated experimental data for the conventional DBI algorithm using datasets provided by the Bioacoustics Research Lab at the University of Illinois at Urbana-Champaign (https://www.brl.uiuc.edu/Projects/inverse_scattering_study .php). After ensuring the accuracy of the DBI model, we proceeded to enhance it by incorporating the proposed techniques, relying solely on simulation data. While experimental validation remains essential in demonstrating the real-world effectiveness of ultrasound imaging methods like DBI, simulation-based development plays a critical role in refining methodologies. Through simulations, we can systematically evaluate and optimize various techniques within the DBI framework before moving toward experimental verification. This approach helps establish a strong foundation for future experimental studies and facilitates the eventual application of DBI advancements in clinical settings.

3. INTERPOLATION TECHNIQUE UTILIZED IN THE DBI METHOD

3.1. Interpolation-based DBI method

The application of the interpolation technique, specifically the multi-resolution (MR) method, to Diffraction Tomography-based Ultrasound Imaging (DTUI) is well documented in existing studies (Yen et al., 2018; Tran-Duc et al., 2011; Huy et al., 2014; Huy et al., 2019). Nonetheless, an important parameter (r, representing the ratio between the number of measurements and the number of variables) has yet to be fully explored with respect to its impact on the imaging quality of the entire system. It is worth emphasizing that an unsuitable selection of this parameter could potentially compromise the efficacy of the interpolation approach, leading to inferior reconstruction performance compared to that of the standard DBI approach.

The complexity of an imaging system is determined by various factors, including the number of pixels of size $N \times N$, the number of transmitters N_t and receivers N_r, and the number of iterations N_{iter}. In this work, we hold the number of pixels N^2 constant and focus on determining the number of N_{iter} iterations. Specifically, we calculate the number of iterations implemented with a raw meshed integration area of size $N_1 \times N_1$, denoted as NN_1, and the number of iterations implemented with a meshed integration area of size $N_2 \times N_2$ is given by $NN_2 = N_{iter} - NN_1$. The effectiveness of the reconstruction performance is evaluated using the Relative Residual Error (RRE) metric. A complete depiction of the MR-DBI algorithm is presented in **Algorithm 2.**

Algorithm 2. Proposed MR-DBI

1. Run Steps 1-2 in Algorithm 1 with the size of $N_1 \times N_1$.
For $n = 1$ to N_{N_1}, do
2. Run Steps 3-7 in Algorithm 1.
End For
3. Interpolate $Z_{N_1}^{(n)}$ to achieve $Z_{N_2}^{(0)}$
4. Set $\overline{\Psi}_0 = \overline{\Psi}_{N_2}^{inc}$ **For** $n = N_{N_1}+1$ to N_{iter}, do
5. Run Steps 3-7 in Algorithm 1.
6. Calculate the RRE using (11)
End for

RRE is investigated by varying N_{N_1} as shown in **Algorithm 3**:

For $N_{N_1} = 1$ to $(N_{iter} - 1)$, do 1. Proceed to **Algorithm 1**. 2. Recover the calculated RRE. **End For** 3. Determine the optimal N_{N_1} value that corresponds to the lowest RRE value.

3.2. Influence of number of transducers on the image reconstruction quality in the DBI and MR-DBI.

Simulation parameters: Frequency f = 0.64MHz; $N_t = 11$; $N_r = 22$, $N_{sum} = 8$; $N_{i1} = 1$; $N_{i2} = N_{sum} - N_{i1}$; $N_{N_1} = 11$; $N_{N_2} = 22$. The diameter of the scattering area is 10mm, while the sound contrast is set to 3%. Additionally, we introduce Gaussian noise at 10% intensity. The distance from the transmitters and receivers to the center of the object is 100mm.

The DBI approach's primary drawback is that the divergence problem arises in situations where

$$\Delta\varphi = 2\omega\left(\frac{1}{c} - \frac{1}{c_0}\right)R, \qquad (12)$$

is the excess phase term represents the change in phase between the incident field and the wave that propagates through the object. Neglecting this term can result in significant artifacts near the center of the object, especially when high sound contrast and large inhomogeneities are present (Lavarello & Oelze, 2009). To achieve an artifact-free reconstruction, it is imperative that $\Delta\varphi < \pi$. In our work, we set the excess phase parameter 0.094π ($\Delta\varphi = 0.094\pi$ based on the values $c_0 = 1540$ m/s, $c = 1493.8$ m/s, $R = 3{,}65$ mm, and f=0.64MHz, which ensures compliance with the Born condition for convergent reconstruction ($\Delta\varphi < \pi$).

To satisfy the Born condition, we selected the initial frequency of the incident signal f= 0.64MHz in our simulation. In the women's breast environment, ultrasonic wave propagation speeds range from 1350 m/s to 1600 m/s (with a background median speed of around 1484 m/s) (Greenleaf & Bahn, 1981). As a result, the variation in propagation speed in women's breasts is between 0% and 15.6%. 3% as the difference percentage in this work is therefore appropriate. The number of pixels used in the simulation is N_1 for NN_1 iterations, and N_2 for NN_2 iterations, resulting in a total of $N_1 \times N_1 = 11 \times 11 = 121$ variables for NN1 iterations and $N_2 \times N_2 = 22 \times 22 = 484$ variables for NN_2 iterations. The number of transmitters and receivers, denoted by N_t and N_r, respectively, is chosen based on the specific simulation scenario, with the total number of measurements equal to $N_t \times N_r$.

3.2.1. Consider $N_T = N_R$

The reconstructed results for the case of $N_T = N_R$ are presented in Table 1. The results demonstrate that when the number of measurements is small (i.e., $r \leq 0.55$), the MR-DBI outperforms the DBI. However, when the number of measurements is large (i.e., $r > 0.55$), the MR-DBI's performance is inferior to that of the DBI. This indicates that the MR-DBI is only significant when the scattering domain is sparse. The optimal outcome occurs when $N_T = N_R = 12$ (i.e. $r = 0.53$), and the MR-DBI yields a minimum error of 0.7883.

3.2.2. Consider $N_T > N_R$ ($N_T = 2N_R$)

Table 2 presents the reconstructed results when $N_T > N_R$. It is evident that when the number of measurements is small (i.e., $r \leq 1.0$), the MR-DBI outperforms the DBI. However, for large numbers of measurements (i.e., $r > 1.0$), the MR-DBI exhibits inferior performance compared to the DBI. This implies that the MR-DBI is only meaningful in the presence of a sparse scattering domain. The optimal outcome is achieved for $N_T = 22$; $N_R = 12$ (i.e. $r = 0.97$, yielding the smallest error of 0.1668 for the MR-DBI.

3.2.3. Consider $N_T < N_R$ ($N_R = 2N_T$)

The reconstructed outcomes under the scenario of $N_T < N_R$ are presented in Table 3. Notably, our findings indicate that the MR-DBI exhibits superior performance compared to the DBI as the number of measurements is small (i.e., . $r \leq 1.0$). However, for large numbers of measurements (i.e., $r > 1.0$), the MR-DBI's performance deteriorates compared to that of the DBI. This implies that the MR-DBI is only effective in case of a sparse scattering domain. The optimal performance is achieved when $N_T = 12$; $N_R = 22$ (i.e. $r = 0.97$), with the smallest MR-DBI error of 0.1654.

To summarize, Tables 1, 2, and 3 indicate that the MR-DBI is only relevant when the value of r is less than 1.0. Furthermore, the optimal scenario for reconstruction is observed when the number of receivers, N_r, is twice the number of transmitters, N_t.

Table 1. **Normalized errors of the DBI and MR-DBI in the situation of $N_T = N_R$**

Different number of transmitters and receivers		Normalized errors from one-to-eight iterations	Total runtime
$N_T = N_R = 6$	DBI	1.1425 1.1294 1.1241 1.1240 1.1240 1.1239 1.1239 1.1239	27.185685
	MR-DBI	1.1890 0.9432 0.7841 0.7123 0.7052 0.7046 0.7045 0.7045	27.394105
$N_T = N_R = 8$	DBI	1.1115 0.9987 0.9778 0.9749 0.9744 0.9744 0.9743 0.9743	32.890772
	MR-DBI	0.9746 1.9515 1.1066 0.7778 0.7194 0.7098 0.7081 0.7077	35.109292
$N_T = N_R = 10$	DBI	1.1315 1.0717 1.0154 1.0039 1.0005 0.9992 0.9985 0.9982	40.663465
	MR-DBI	1.1755 2.3464 1.2545 1.0032 0.9127 0.8665 0.8398 0.8237	41.173181
$N_T = N_R =$ **12** $N_T x N_R=144$, r=0.53	**DBI**	**0.9751 0.9044 0.8695 0.8605 0.8568 0.8551 0.8542 0.8538**	**47.922455**
	MR-DBI	**1.0615 2.8415 1.5233 1.0982 0.9371 0.8590 0.8149 0.7883**	**43.809723**
$N_T = N_R = 14$	DBI	0.7069 0.6502 0.5987 0.5879 0.5840 0.5821 0.5809 0.5801	53.584532
	MR-DBI	0.7774 4.0380 2.4126 1.6840 1.3354 1.1428 1.0246 0.9507	57.974882
$N_T = N_R = 16$	DBI	0.4741 0.3196 0.2466 0.2322 0.2259 0.2216 0.2184 0.2158	65.557087
	MR-DBI	0.4857 5.1182 3.4697 2.6304 2.1919 1.9180 1.7225 1.5745	66.505106

continued on following page

Table 1. Continued

Different number of transmitters and receivers		Normalized errors from one-to-eight iterations	Total runtime
$N_T = N_R = 18$	DBI	0.4911 0.3644 0.2867 0.2678 0.2575 0.2508 0.2458 0.2418	76.620161
	MR-DBI	0.5056 0.5618 0.4288 0.3404 0.2887 0.2552 0.2319 0.2147	72.901932
$N_T = N_R = 20$	DBI	0.2954 0.2051 0.1876 0.1756 0.1704 0.1668 0.1641 0.1620	94.407173
	MR-DBI	0.2388 0.7838 0.6164 0.4902 0.4136 0.3598 0.3202 0.2902	100.612536
$N_T = N_R = 22$	DBI	0.3177 0.2170 0.1922 0.1751 0.1658 0.1592 0.1546 0.1512	111.652342
	MR-DBI	0.2859 0.7431 0.5736 0.4490 0.3733 0.3224 0.2862 0.2597	120.522988
$N_T = N_R = 24$	DBI	0.2536 0.1941 0.1736 0.1624 0.1548 0.1499 0.1462 0.1433	130.047672
	MR-DBI	0.1843 0.7409 0.6056 0.4692 0.3949 0.3475 0.3135 0.2872	137.976071
$N_T = N_R = 26$	DBI	0.2658 0.1971 0.1790 0.1621 0.1525 0.1464 0.1420 0.1389	148.980734
	MR-DBI	0.2062 0.7383 0.6361 0.4955 0.4098 0.3525 0.3122 0.2822	161.676352
$N_T = N_R = 28$	DBI	0.2599 0.1960 0.1748 0.1581 0.1485 0.1425 0.1383 0.1353	168.739083
	MR-DBI	0.1891 0.7552 0.6053 0.4665 0.3860 0.3329 0.2961 0.2694	184.583668
$N_T = N_R = 30$	DBI	0.2488 0.1952 0.1692 0.1538 0.1450 0.1393 0.1355 0.1326	640.904832
	MR-DBI	0.1733 0.8846 0.6355 0.4879 0.3989 0.3401 0.2996 0.2697	210.240263

Table 2. **Normalized errors of the DBI and MR-DBI in the situation of $N_T > N_R$**

Different number of transmitters and receivers		**Normalized errors from one-to-eight iterations**	**Total runtime**
$N_R = 12; N_T = 14$	DBI	0.4434 0.3218 0.2996 0.2970 0.2966 0.2964 0.2964 0.2964	54.821645
	MR-DBI	0.4825 0.3271 0.2446 0.2320 0.2323 0.2326 0.2328 0.2329	46.915194
$N_R = 12; N_T = 16$	DBI	0.4568 0.3313 0.2989 0.2955 0.2948 0.2946 0.2946 0.2946	67.127392
	MR-DBI	0.5020 0.2886 0.2207 0.2049 0.2034 0.2033 0.2033 0.2033	47.864321
$N_R = 12; N_T = 18$	DBI	0.4297 0.2801 0.2464 0.2397 0.2375 0.2365 0.2360 0.2357	59.296794
	MR-DBI	0.4198 0.3151 0.2240 0.1929 0.1903 0.1900 0.1900 0.1901	54.958475
$N_R = 12; N_T = 20$	DBI	0.4230 0.2754 0.2428 0.2354 0.2320 0.2298 0.2285 0.2275	67.596661
	MR-DBI	0.4814 0.3301 0.2161 0.1805 0.1737 0.1716 0.1708 0.1705	55.043968
$N_R = 12; N_T = 22$ $N_T x N_R = 264$, r=0.97	**DBI**	**0.4293 0.2725 0.1996 0.1870 0.1811 0.1780 0.1760 0.1746**	**97.148919**
	MR-DBI	**0.4533 0.3394 0.2239 0.1776 0.1708 0.1688 0.1676 0.1668**	**58.610749**
$N_R = 22; N_T = 26$	DBI	0.4078 0.1800 0.0828 0.0666 0.0611 0.0577 0.0552 0.0529	201.430871
	MR-DBI	0.4065 0.4937 0.3564 0.2661 0.2233 0.1968 0.1784 0.1645	113.768921
$N_R = 22; N_T = 30$	DBI	0.4020 0.1830 0.0831 0.0645 0.0592 0.0559 0.0531 0.0507	159.909376
	MR-DBI	0.3885 0.5410 0.4024 0.3048 0.2556 0.2231 0.1993 0.1812	136.049329
$N_R = 22; N_T = 34$	DBI	0.4032 0.1701 0.0837 0.0622 0.0564 0.0526 0.0494 0.0466	159.673462
	MR-DBI	0.3904 0.5130 0.3669 0.2625 0.2162 0.1871 0.1665 0.1508	155.827023
$N_R = 22; N_T = 38$	DBI	0.4038 0.1701 0.0822 0.0611 0.0553 0.0512 0.0477 0.0447	204.449995
	MR-DBI	0.3886 0.5700 0.4150 0.3037 0.2470 0.2125 0.1881 0.1697	164.678051
$N_R = 22; N_T = 42$	DBI	0.4033 0.1668 0.0807 0.0596 0.0533 0.0487 0.0448 0.0415	300.550899
	MR-DBI	0.3842 0.5340 0.4231 0.3045 0.2452 0.2098 0.1860 0.1692	243.442987

continued on following page

Table 2. Continued

Different number of transmitters and receivers		**Normalized errors from one-to-eight iterations**	**Total runtime**
$N_R = 22; N_T = 44$	DBI	0.4033 0.1668 0.0807 0.0596 0.0533 0.0487 0.0448 0.0415	207.872891
	MR-DBI	0.3842 0.5436 0.4392 0.3199 0.2583 0.2180 0.1892 0.1682	193.711118

Table 3. **Normalized errors of the DBI and MR-DBI in the situation of** $N_T < N_R$

Different number of transmitters and receivers		**Normalized errors from one-to-eight iterations**	**Total runtime**
$N_T = 12; N_R = 14$	DBI	0.4504 0.3249 0.2943 0.2903 0.2891 0.2887 0.2885 0.2884	85.916649
	MR-DBI	0.5146 0.3153 0.2424 0.2332 0.2331 0.2332 0.2332 0.2332	52.919105
$N_T = 12; N_R = 16$	DBI	0.4540 0.3541 0.3292 0.3251 0.3240 0.3237 0.3236 0.3236	71.862117
	MR-DBI	0.4833 0.3022 0.2361 0.2223 0.2220 0.2222 0.2223 0.2224	47.607924
$N_T = 12; N_R = 18$	DBI	0.4328 0.3028 0.2524 0.2438 0.2410 0.2399 0.2392 0.2389	62.858595
	MR-DBI	0.4545 0.3869 0.2494 0.2026 0.1930 0.1902 0.1889 0.1882	59.101650
$N_T = 12; N_R = 20$	DBI	0.4311 0.2982 0.2626 0.2579 0.2559 0.2549 0.2544 0.2540	63.310584
	MR-DBI	0.4478 0.3256 0.2156 0.1774 0.1716 0.1702 0.1697 0.1695	58.653183
$\mathbf{N_T= 12; N_R= 22}$ $N_T x N_R = 264$, r=0.97	**DBI**	**0.4293 0.2678 0.2035 0.1918 0.1873 0.1848 0.1833 0.1824**	**65.988784**
	MR-DBI	**0.4558 0.3881 0.2397 0.1834 0.1718 0.1682 0.1664 0.1654**	**59.617548**
$N_T = 22; N_R = 26$	DBI	0.4047 0.1980 0.0898 0.0680 0.0612 0.0567 0.0531 0.0502	127.410144
	MR-DBI	0.3926 0.5027 0.3865 0.2937 0.2457 0.2163 0.1962 0.1813	120.905953
$N_T = 22; N_R = 30$	DBI	0.4010 0.1921 0.0876 0.0621 0.0553 0.0505 0.0467 0.0435	139.577423
	MR-DBI	0.3764 0.5473 0.3941 0.2843 0.2319 0.2018 0.1816 0.1666	133.775860

continued on following page

Table 3. Continued

Different number of transmitters and receivers		Normalized errors from one-to-eight iterations	Total runtime
$N_T = 22; N_R = 34$	DBI	0.4023 0.1919 0.0866 0.0602 0.0531 0.0482 0.0443 0.0411	159.059236
	MR-DBI	0.3831 0.5123 0.4228 0.3141 0.2592 0.2244 0.1995 0.1805	146.832556
$N_T = 22; N_R = 38$	DBI	0.4028 0.1920 0.0888 0.0591 0.0517 0.0464 0.0421 0.0385	174.380491
	MR-DBI	0.3811 0.5209 0.4154 0.3084 0.2548 0.2206 0.1967 0.1787	164.110652
$N_T = 22; N_R = 42$	DBI	0.4021 0.1897 0.0857 0.0570 0.0489 0.0432 0.0388 0.0353	233.711521
	MR-DBI	0.3739 0.5712 0.4569 0.3412 0.2752 0.2323 0.2020 0.1791	189.283227
$N_T = 22; N_R = 44$	DBI	0.4021 0.1897 0.0857 0.0570 0.0489 0.0432 0.0388 0.0353	214.085054
	MR-DBI	0.3739 0.5595 0.4380 0.3244 0.2653 0.2270 0.2009 0.1814	214.340850

3.3. Find the best value of N_{N_1}

The simulation parameters are presented in Table 4, while Table 5 displays the correlation between the number of measurements and variables for each scenario.

Table 4. Scenario simulation parameters

Scenarios Parameters	1	2	3	4	5
N_1	11	13	15	18	21
N_2	22	26	30	36	42
Other parameters	Frequency f = 0.64MHz; $N_t = 11$; $N_r = 22$, $N_{sum} = 8$; N_{i1}; $N_{i2} = N_{sum} - N_{i1}$; Scattering area diameter = 10mm; Sound contrast 3%; Gaussian noise 10%; Distance between the transmitters and receivers and the object's center are 100mm and 100mm, respectively.				

Table 5. The correlation between the quantity of measurements and scenario variables

Scenarios Parameters	**1**	**2**	**3**	**4**	**5**
Measurements ($N_t \times N_r$)	242	242	242	242	242
Variables $[(N_1+N_2)/2]^2$	272.25	380.25	506.25	729	992.25
Measurements/ Variables	0.88	0.64	0.48	0.33	0.24
N_{N1}	$N_{N1} = 2$		$N_{N1} = 3$		

Figure 2. Normalized errors of the MR-DBI through iterations corresponding to distinct N_{i1} *values in the first scenario (The smallest error is obtained when* $N_{N1} = 2$*)*

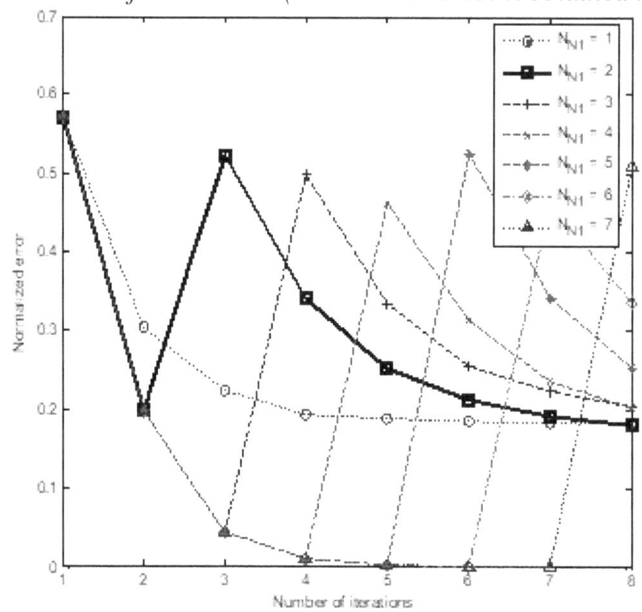

Figure 3. Normalized errors of the MR-DBI through iterations corresponding to different N_{i1} *values in the second scenario (The smallest error is obtained when* $N_{N1} = 2$*)*

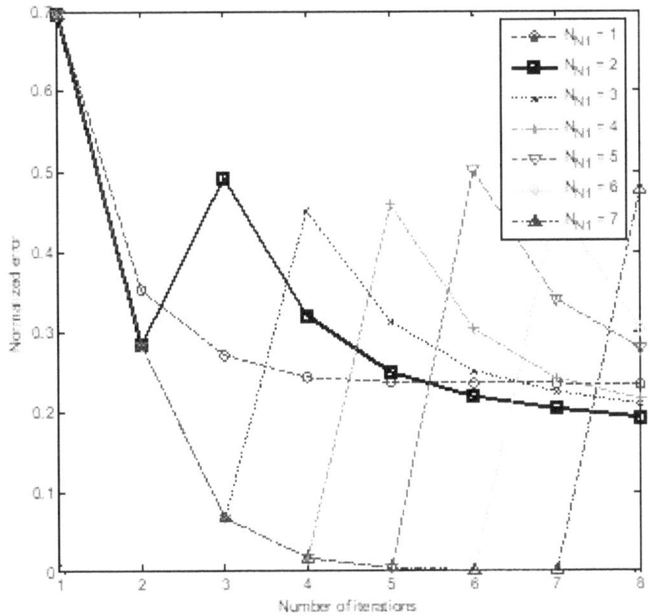

Figure 4. The MR-DBI faults are normalized over iterations corresponding to distinct N_{i1} *values in the third situation (The smallest error is obtained when* $N_{N1} = 3$*)*

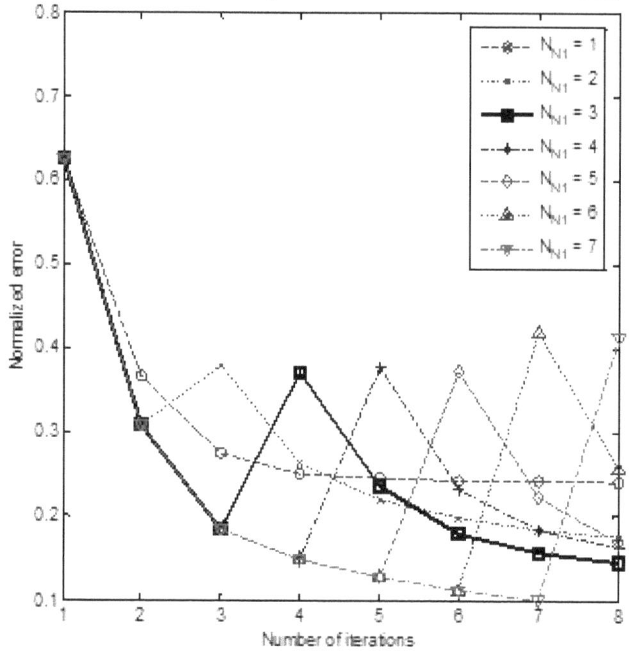

Figure 5. The MR-DBI faults are normalized over iterations corresponding to distinct N_{i1} values in the fourth situation (The smallest error is obtained when N_{N1} = 3)

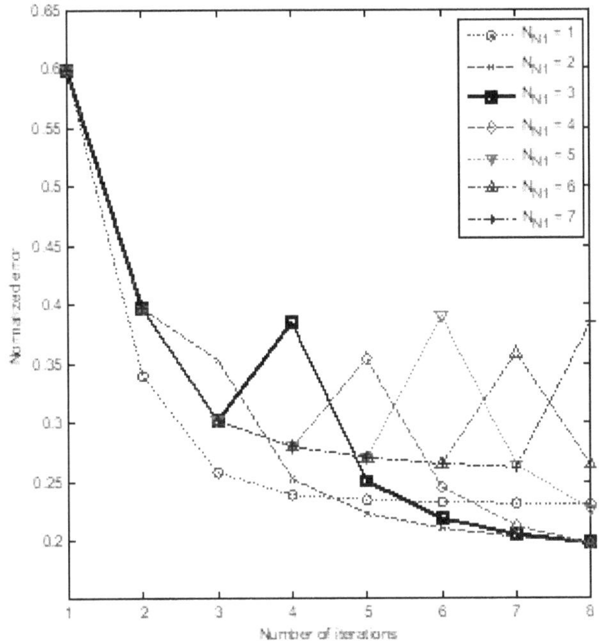

Figure 6. The MR-DBI faults are normalized over iterations corresponding to distinct N_{i1} values in the fifth situation (The smallest error is obtained when $N_{N1} = 3$)

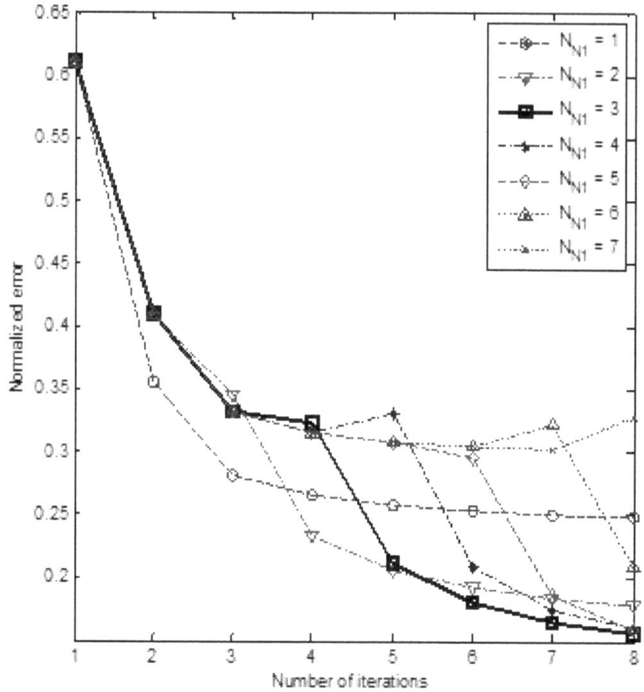

Figure 7. MR-DBI error normalization through iterations corresponding to different N_T *and* N_R

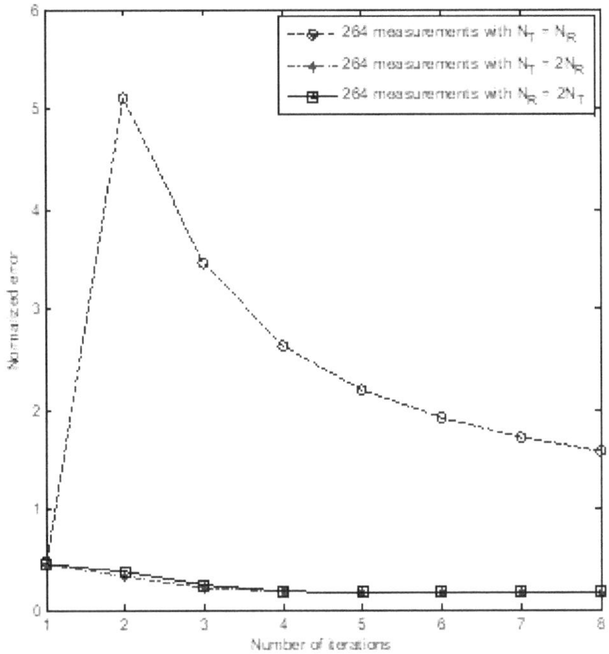

3.4. Simulation of DBI and MR-DBI

Figure 8 displays the ideal object function, denoted by O(r), which is positioned at the center of the meshing area.

Figure 8. Ideal object function (N = 22)

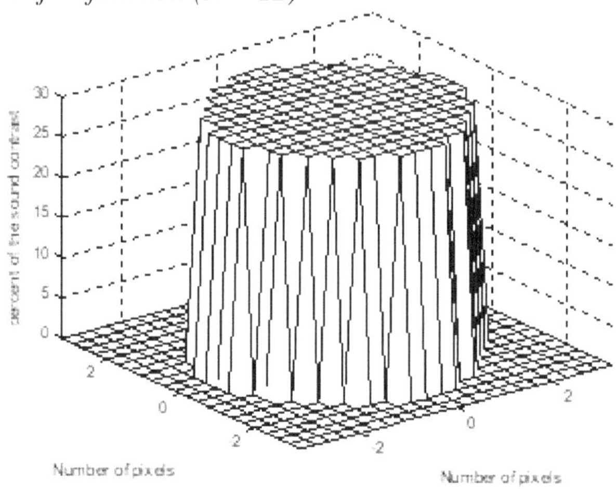

In our recent study (Huy et al., 2019), a chart of the total runtime for different methods after N_{sum} iterations is presented. By applying the interpolation approach, the simulation results demonstrate a reduction of 33% in the total imaging runtime. The numerical simulation program was executed on a PC equipped with an Intel Core i3 processor and 2 GB RAM, utilizing Matlab. Based on the simulated dataset, the computational complexity of the conventional DBI is calculated as $O\left(N_{sum}N_tN_rN_{N2}^2\right) = 8\times11\times22\times22^2 = 937024$ calculations. The computational cost of our suggested technique is merely $O\left(N_{i1}N_tN_rN_{N1}^2\right) + O\left(N_{i2}N_tN_rN_{N2}^2\right) = 2\times11\times22\times11^2 + 6\times11\times22\times22^2 = 58564 + 702768 = 761332$ calculations. This represents a reduction of 175692 calculations, equivalent to approximately 18.75% when compared to the conventional method.

Aside from the benefits, this work has the following limitations: (a) A nearest-neighbor interpolation, which replaces each pixel with four pixels of the same color, is one of the simplest ways to double its size. The final image is bigger than the original and has all of the original information. The closest neighbor method is the simplest and fastest method of interpolation. As a result, this technique is only appropriate for applications requiring rapid imaging and processing time, such as real-time imaging or online data transfer. This technology can blur pictures for high-resolution and exact imaging requirements; (b) This study is only implemented using simulation data; imaging with experimental data requires additional investigation; and (c) The closest neighbor interpolation method employed is still relatively simple. To increase the impact of the research findings, the number of iterations for NN_1 should be surveyed in various interpolation algorithms such as bilinear, bicubic, splines, and so on. (d) The modeling is simple, with the tumor in the center of the circle; the work may be enhanced by changing the size of the tumor and increasing the number of individuals in the region of interest.

4. DUAL-FREQUENCY TECHNIQUE UTILIZED IN THE DBI METHOD

4.1. Dual-frequency-based DBI method

In this study, we denote the number of iterations performed using the function f_1 as N_{f_1}. Similarly, the number of iterations performed using f_2 is defined as $N_{f_2} = N_{iter} - N_{f_1}$.

Algorithm 4. presents a complete description of DF-DBI.

Algorithm 4. Suggested DF-DBI

1. Run Steps 1-2 in Algorithm 1.
For n =1 to N_{f_1}, do
2. Run Steps 3-7 in Algorithm 1 with lower frequency f_1.
End For
For n = N_{f_1}+1 to N_{iter}, do
3. Run Steps 3-7 in Algorithm 1 with higher frequency f_2.
4. Calculate the RRE using (11)
End for

After **Algorithm 4**, the investigation of RRE involves variation of N_{f_1}, which is outlined in the following procedure depicted in **Algorithm 5:**

1. **For** N_{f_1} = 1 to (N_{iter} − 1), do 2. Go to **Algorithm 4** 3. Restore the computed RRE. 4. **End For** 5. Determine the best choice of N_{f_1} corresponding to the minimum value of RRE.

The simulation parameters for the experiments are presented in Table 6, which contains the values of the number of transmitters (N_t) and receivers (N_r) for each scenario. The objective function involves the estimation of $N \times N$ variables with the help of $N_t \times N_r$ measurements. The first scenario corresponds to large values of N_t and N_r, the second scenario deals with small values of N_t and N_r, while the third and fourth scenarios involve moderate values of N_t and N_r.

Table 7 presents the correlation between the number of measurements and variables for each situation. In the first situation, the number of equations surpasses the number of variables, indicating an over-determined problem resulting from a high number of transmitters and receivers. Thus, this condition leads to a significant increase in the computational complexity of the system. On the other hand, scenarios two, three, and four show fewer equations than variables, which defines an ill-conditioned problem with a small to moderate number of transmitters and receivers. Consequently, this situation is more appropriate for practical applications.

Table 6. Simulation parameters of scenarios

vvvvvvvvvvvvvvvvvvvScenarios Parameters	1	2	3	4
N_t	44	15	22	27
N_r	22	7	11	14
Other parameters	Frequencies f_1 = 1MHz, f_2 = 2MHz; N = 22, N_{iter} = 8; Scattering area diameter 10mm; Wave propagation speed difference 30%; Gaussian noise 10%; Distance between the transmitters and receivers and the object's center are 50mm and 60mm, respectively.			

Table 7. The correlation between the quantity of measurements and scenario variables

vvvvvvvvvvvvvvvvvvvvvvvvvvvvScenarios Parameters	1	2	3	4
Variables (N×N)	484	484	484	484
Measurements (N_t×N_r)	968	105	242	378
Measurements/ Variables	2	0.22	0.5	0.78

4.2. Determine the optimal value for N_{f_1}

The figures in Figs. 3, 4, 5, and 6 demonstrate the normalized errors of the DF-DBI algorithm through different iterations corresponding to various N_{f_1} values for the first, second, third, and fourth scenarios, respectively. Table 8 presents the errors of scenarios corresponding to each value of N_{f_1} after N_{iter} iterations. Based on the simulation results, the value of N_{f_1} is dependent on the number of N_t and N_r. Specifically, if N_t and N_r are small, N_{f_1} is large, and the result depends solely on f_1, as indicated by Figure 10 (N_{f_1} = 6). On the other hand, when N_t and N_r are large, the result depends only on f_2, as shown in Figure 9 (N_{f_1} = 1). This finding suggests that the dual-frequency approach does not provide a better performance than the single frequency one. In fact, when N_t×N_r is greater than N×N, the number of equations is larger than the number of variables, ensuring convergence of the DF-DBI algorithm. Under such circumstances, only the higher frequency f_2 can offer a better performance. Conversely, if N_t and N_r are such that N_t×N_r is smaller than N×N, the number of equations becomes less than the number of variables, and the DF-DBI algorithm convergence can be achieved using only the lower frequency f_1. For moderate values of N_t and N_r (i.e., third and fourth scenarios), N_{f_1} can be selected to exploit both f_1 and f_2. It is evident from Figs. 11 and 12 ($N_{f_1} = 4$) that the DF-DBI algorithm is especially meaningful in moderate N_t and N_r scenarios. Therefore, we choose scenario 3 for further investigation, where the number of transmitters and receivers is half the number of variables.

Table 8. Situation errors for each value of N_{f_1} *after a total of 8 iterations*

N_{f_1} **Errors**	**1**	**2**	**3**	**4**	**5**	**6**	**7**
Situation 1	**0.0049**	0.0081	0.0098	0.0169	0.0284	0.0514	0.1537
Situation 2	0.3420	0.3266	0.3268	0.3273	0.3260	**0.3248**	0.3364
Situation 3	0.2510	0.1980	0.1847	**0.1800**	0.1810	0.1863	0.2402
Situation 4	0.1544	0.1061	0.0984	**0.0970**	0.1078	0.1235	0.2013

Figure 9. Normalized DF-DBI errors during iterations corresponding to varied N_{f_1} *values in the first case*

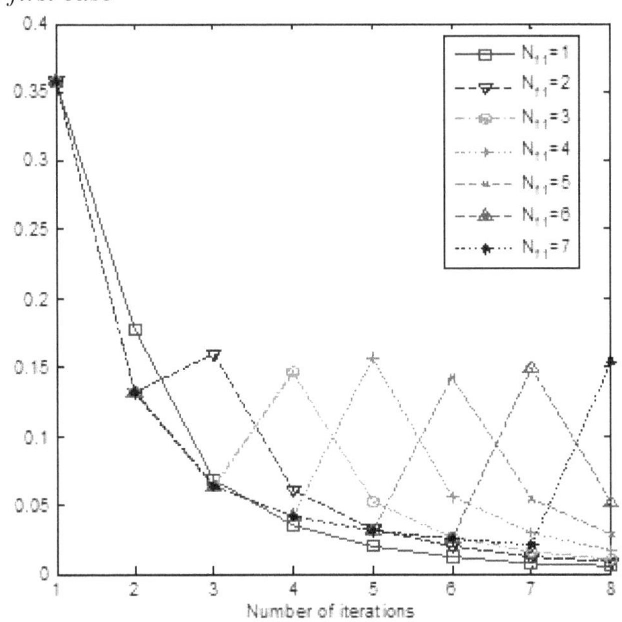

Figure 10. Normalized DF-DBI errors during iterations corresponding to varied N_{f_1} *values in the second case*

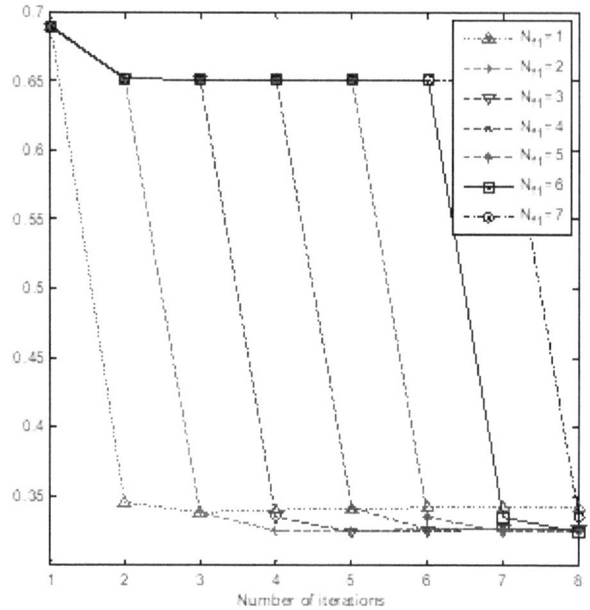

Figure 11. Normalized DF-DBI errors during iterations corresponding to varied N_{f_l} *values in the third case*

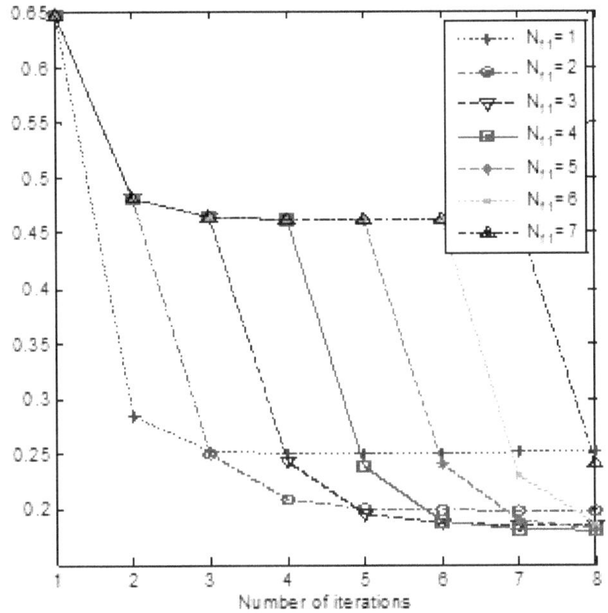

Figure 12. Normalized DF-DBI errors during iterations corresponding to varied N_{f_1} values in the fourth case

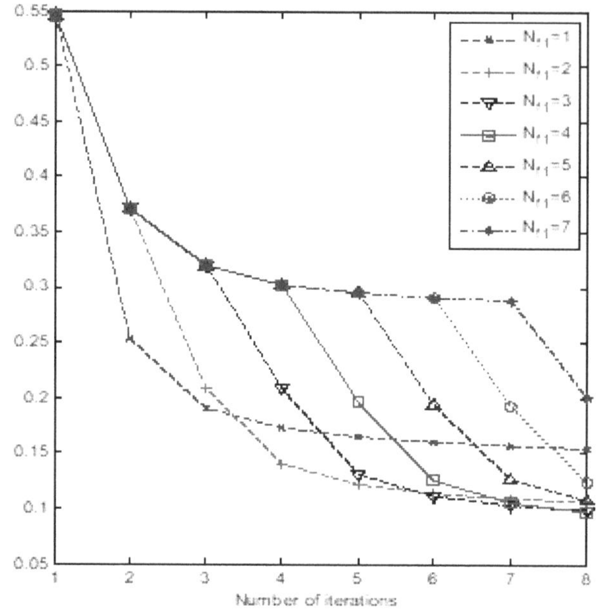

4.3. Simulation of DBI and DF-DBI

Scenario 3 involves setting N_{iter} to 8, with the optimal iteration value for the first frequency found to be $N_{f_1} = N_{iter}/2$ (which is equal to 4). Various simulations were conducted with different input parameters (including moderate values of N_{iter}), and the results confirmed that $N_{iter}/2$ was indeed the best iteration value for the first frequency. This observation is not unexpected, given that when working mainly with f_1 (with N_{f_1} close to N_{iter} and N_{f_2} close to 1), the algorithm quickly converges but provides low precision. Conversely, the algorithm gives high precision with low convergence speed when working primarily with f_2 (with N_{f_2} close to N_{iter} and N_{f_1} close to 1). Therefore, a reasonable choice for N_{f_1} would be to strike a balance between N_{f_1} and N_{f_2}, i.e., $N_{f_1} = N_{iter}/2$.

To demonstrate the efficacy of the suggested technique, Figs. 13 and 15 show the performance of the normalized errors of three alternative ways (using f_1, using f_2, and combining f_1 and f_2) in two situations 3 and 4. When compared to the traditional DBI approach, which employs a single frequency, it can be shown that the normalized error is reduced by 67.6%. This underscores the fact that the two-frequencies compounding method may undoubtedly benefit from both lower and higher frequencies. It provides a higher rate of convergence and a lower normalized error. Figures 14 and 16 display the radial profiles of reconstruction for two scenarios (3 and 4), including the ideal profile (solid curve) and the final DBI reconstructions obtained using f_1, f_2, and both f_1 and f_2. It is evident from these figures that the proposed scheme utilizing both f_1 and f_2 yields the most optimal performance.

Figure 13. DF - DBI and DBI error comparison after N_{iter} repetitions (Situation 3)

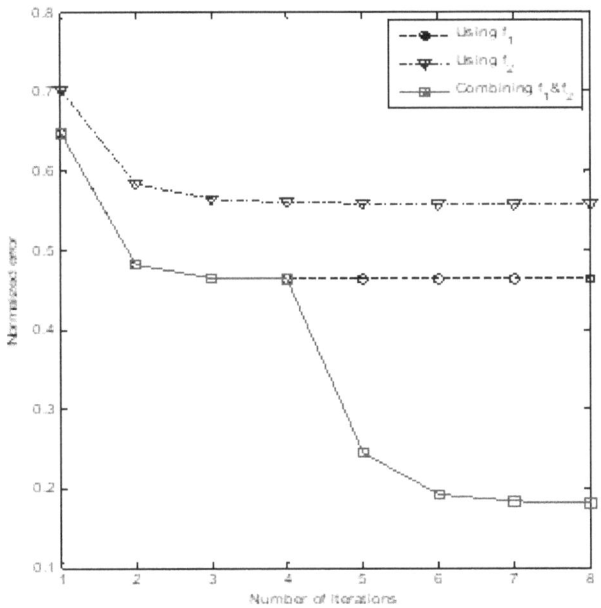

Figure 14. Reconstruction of the profiles of the DF-DBI and the DBI after N iterations (Situation 3)

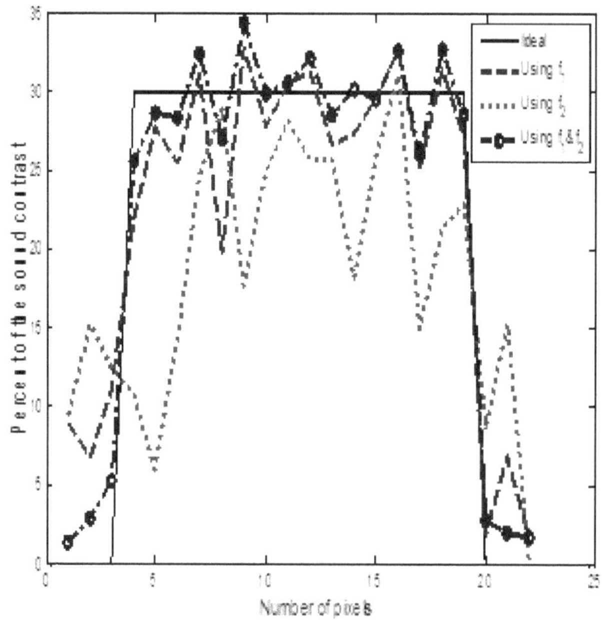

Figure 15. DF - DBI and DBI error comparison after N_{iter} repetitions (Situation 4)

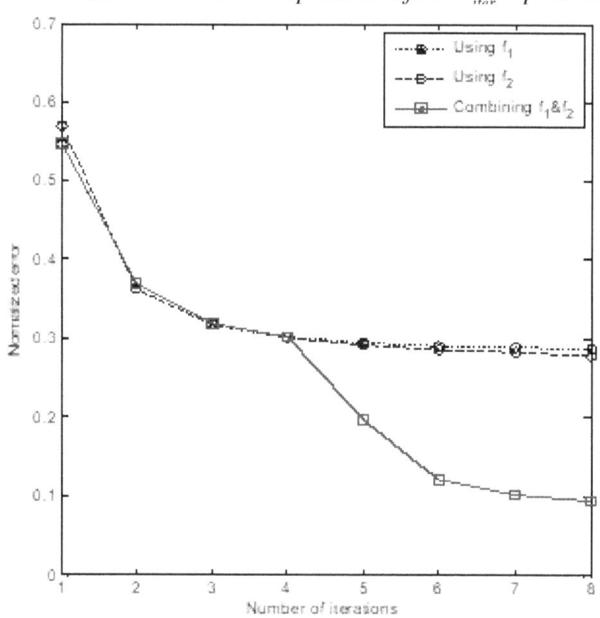

Figure 16. Reconstruction of the profiles of the DF-DBI and the DBI after N iterations (Situation 4)

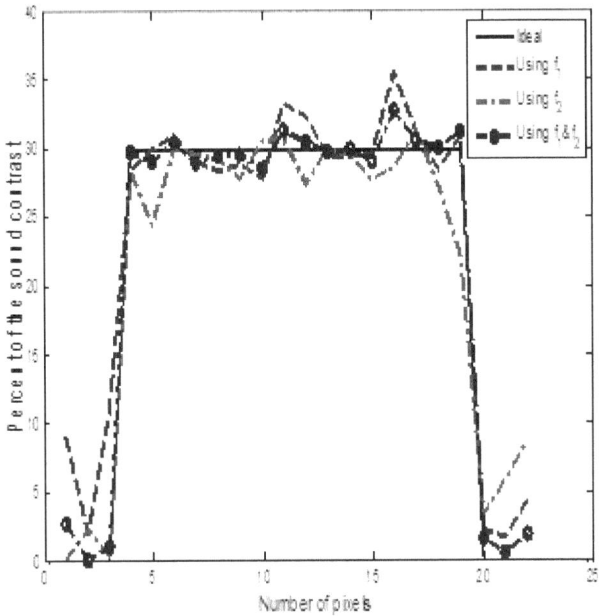

5. THE DBI APPROACH EMPLOYS A MULTI-RESOLUTION AND MULTI-FREQUENCY TECHNOLOGY

5.1. DBI Approach with Multiple Resolutions

The multi-resolution DBI method employed nearest neighbor interpolation. The implementation procedure of the one-step multi-resolution DBI (one-step MR-DBI) was as follows:

$N_{11} \times N_{11}$ $N_{22} \times N_{22}$

The variable N_{N11} represents the quantity of iterations executed on a raw mesh with an integrated area of $N_{11} \times N_{11}$ size. Consequently, the number of iterations performed on a mesh with dimensions of $N_{22} \times N_{22}$ is expressed by $N_{N22} = N_{sum} - N_{N11}$.

The four-step, multi-resolution DBI implementation approach was as follows:

$N_1 \times N_1$ $N_2 \times N_2$ $N_3 \times N_3$ $N_4 \times N_4$

The quantities of iterations carried out with mesh integrated areas of sizes $N_1 \times N_1$, $N_2 \times N_2$, $N_3 \times N_3$, and $N_4 \times N_4$ are respectively represented by the variables N_{N1}, N_{N2}, N_{N3}, and N_{N4}.

The following simulation parameters were utilized: frequency, $f = 0.64$ MHz, $N_t = 11$, $N_r = 22$, $N_{sum} = 8$, $N_{N11} = 17$, $N_{N22} = 33$, $N_1 = 5$, $N_2 = 9$, $N_3 = 17$, $N_4 = 33$, $N_{N1} = 1$, $N_{N2} = 1$, $N_{N3} = 1$, $N_{N4} = 5$. The scattering area diameter was 7.3 mm, the sound contrast was 30%, the Gaussian noise was 10%, the speed of sound in the background was 1540 m/s, and the distances from the transmitters and receivers to the object's center were 50 and 60 mm, respectively.

In DBI, the precise value of each pixel in the area of interest is computed. The ultrasonic wave will disperse if there is a small-scale heterogeneous medium that is comparable to the incident wavelength, and scattering measurements are collected. DBI may be used to pinpoint an object's precise location and form. Therefore, it is clear that the algorithm's capacity to precisely and efficiently retrieve the object is the main issue. Consequently, in this case, the selected model and parameters were designed based on the goal of creating a better picture recovery method than the current one. As a result, a straightforward round cylinder was chosen as the item to be rebuilt, and its surroundings were described as uniform. The investigation's attention then shifted to finding a way to restore the picture as closely as feasible to the goal function. As a result, the sound contrast was the parameter that was utilized to recreate the ideal object; the characteristics of density and attenuation were ignored. The works (Mojabi & LoVetri, 2015; Ghai et al., 2022) contain a study on the impact of the acoustic density, attenuation, and compressibility profile factors on the resulting pictures.

In Figure 17, the ideal functions representing the ground truth of the objects to be imaged are displayed. The more pixels that are involved, the more variables there are to detect and recover, and so the imaging system gets more complex.

*Figure 17. Ideal functions of the objects of interest in terms of the number of involved pixels: 5 (**a**), 17 (**b**), and 33 (**c**), respectively.*

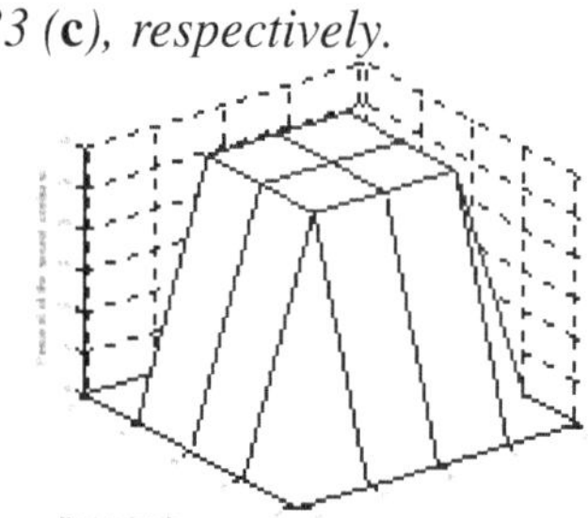

The investigation focused on determining the minimum value of N1 for the first raw meshed integration area, which was found to exhibit superior performance, as presented in Table 9. Analysis of the data in Table 9 indicates that the optimal value of N1 is 5, offering the most favorable results. As a result, $N_1 = 5$ was selected for conducting a more comprehensive simulation.

Table 9. Error after the first iteration using $N_1 \times N_1$ (NoC-no convergence, best value in bold).

N_1	**1**	**2**	**3**	**4**	**5**
Error	NoC	NoC	0.1572	0.1280	**0.1229**
N_1	6	7	8	9	10
Error	0.2082	0.4898	0.5078	0.5278	0.4525
N_1	11	12	13	14	15
Error	0.5725	0.6408	0.6957	0.6304	0.6256
N_1	16	17	18	19	20
Error	0.5991	0.6302	0.5991	0.6215	0.6288

Figure 18 illustrates the error performance of the proposed four-step MR-DBI method relative to the other compared methods. The results indicate a notable reduction in normalized error compared to both the DBI and one-step MR-DBI methods. Further analysis suggests that the RRE improved progressively with each iteration of the one-step MR-DBI method, as estimating the smaller ($N_{11} \times N_{11}$) object with $N_{N11} = 17$ prior to estimating the larger ($N_{22} \times N_{22}$) object with the DBI's $N_{N22} = 33$ yielded better results for the same number of measurements. Therefore, achieving a good estimate in the initial loop of the one-step MR-DBI method led to improved estimation performance overall. Regarding the four-step MR-DBI method, it is apparent that the maximum point was reached at the fourth loop, which is to be expected, as this loop involved restoring the object with the largest ($N_4 \times N_4$) size (the desired value). Since the number of pixels increases with object size, more variables are involved, making estimation more challenging, resulting in a higher RRE. Conversely, in loops one and two, the RRE was relatively low, as the number of variables was small, leading to more accurate estimation.

The total runtime required by the DBI, one-step MR-DBI, and four-step MR-DBI methods after Nsum iterations was computed. Results indicated that the DBI method required the longest imaging time, with a duration of 640.7 seconds. As the interpolation level increased, the imaging time decreased, with the one-step interpolation method requiring 569.8 seconds and the four-step interpolation method requiring 405.5 seconds. This observation is logical since the number of variables, or pixels, represented by ($N_4 \times N_4$), remains constant in each loop with the DBI method. However, with interpolation, the number of pixels gradually increases, ultimately reaching $N_4 \times N_4$, and thus the number of variables in the earlier loops is significantly less than in the DBI method, resulting in a substantial reduction in imaging time.

Figure 18. Normalized error comparison among the DBI, one-step MR-DBI, and four-step MR-DBI methods.

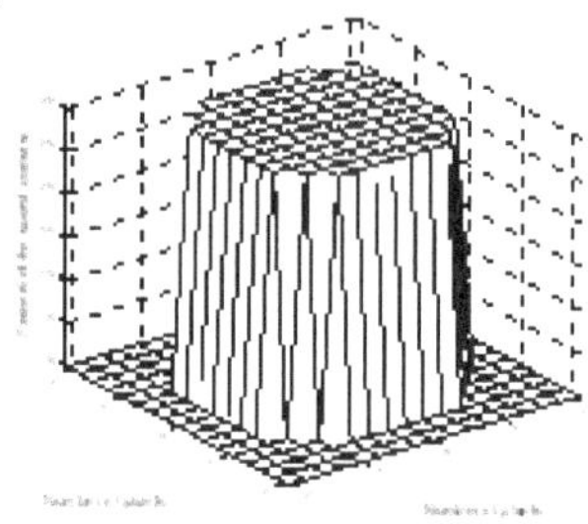

Figure 19. Normalized error comparison between the MR-MF-DBI and four-step MR-DBI methods.

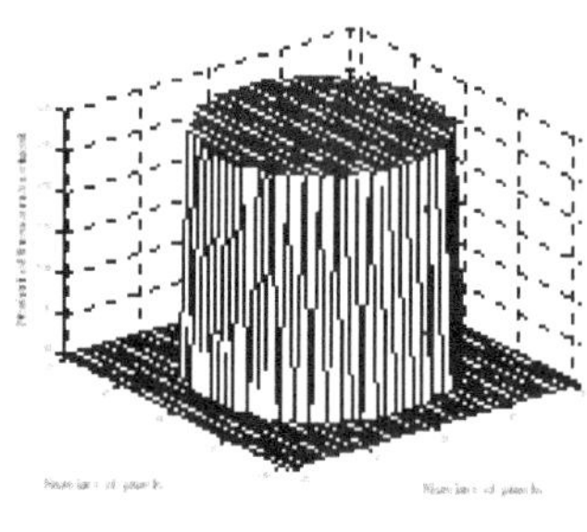

5.2. Multi-Resolution and Multi-Frequency DBI Approach

Undoubtedly, the employment of the multi-resolution and multi-frequency DBI approach (MR-MF-DBI) significantly decreases the image formation time while ensuring accurate estimation of the object of interest in the initial iterations. Hence, for the proposed methodology, a frequency of 0.64 MHz, satisfying the Born approximation condition, was employed in the initial loops, particularly in the first four loops. Subsequently, as the number of variables, i.e., the number of pixels, reached the maximum desired value, the frequency was gradually incremented from the fifth loop onwards, effectively mitigating noise and augmenting the resolution of the recovered image.

The present experiment utilizing the proposed MR-MF-DBI approach was conducted with specific simulation parameters, which include frequency, $f = 0.64$ MHz, $f_1 = 2f$, $f_2 = 3f$, $f_3 = 4f$, $f_4 = 5f$, $N_t = 11$, $N_r = 22$, $N_{sum} = 8$, $N_1 = 5$, $N_2 = 9$, $N_3 = 17$, $N_4 = 33$, $N_{N1} = 1$, $N_{N2} = 1$, $N_{N3} = 1$, $N_{N4} = 5$, $N_{N4f} = 1$, $N_{N4f1} = 1$, $N_{N4f2} = 1$, $N_{N4f3} = 1$, $N_{N4f4} = 1$, along with a scattering area diameter of 7.3 mm, a sound contrast of 30%, Gaussian noise of 10%, and transmitter and receiver distances of 50 mm and 60 mm, respectively, from the center of the object of interest.

Figure 19 illustrates the error performance of the proposed MR-MF-DBI method relative to the MR-DBI method. The results indicate a notable reduction in normalized error compared to the four-step MR-DBI method. Moreover, a reduction in RRE was observed in the fourth loop as frequency increased. This outcome is attributable to the effective correction of noise by increasing frequency and the gradual updating of the image's sound contrast, resulting in improved estimation.

The total runtimes of the MR-MF-DBI and four-step MR-DBI methods after Nsum iterations were computed. Despite the significant reduction in RRE observed with the MR-MF-DBI method relative to the four-step MR-DBI method, the imaging time increased by only 8.9%. This outcome occurred primarily in loops five to eight of the MR-MF-DBI method due to frequency increases and substantial numerical values in the imaging matrix, leading to longer computational times.

The use of the MR-DBI method can lead to improved imaging quality (as depicted in Figure 18) and faster imaging process. However, in order to meet actual application requirements and achieve higher resolution image reconstruction while reducing noise, a multi-frequency technique was employed. The MR-MF-DBI method was found to produce images with lower background noise in the recovered target compared to the MR-DBI method. As suggested in previous studies (Haddadin & Ebbini, 1998; Tijhuis et al., 2001) the multi-frequency technique can further enhance image resolution.

The accuracy of image reconstruction improves with an increase in data collection. In the DBI method, the number of measurements is determined by the product of the number of transmitters and receivers. Figure 20 depicts the quality of the reconstructed image obtained by the proposed MR-MF-DBI method, based on five scenarios: $N_t \times N_r$ equal to: $11 \times 22 = 242$, $13 \times 24 = 312$, $15 \times 26 = 390$, $17 \times 28 = 476$, and $19 \times 30 = 570$ measurements. The normalized error decreased as the number of measurements increased, indicating a rise in imaging quality. This outcome is expected since the number of pixels, or variables, remains constant as the number of measurements increases. Therefore, more data can be collected, resulting in more precise reconstruction of the object of interest.

Figure 21 presents an analysis of the reconstructed image quality using MR-MF-DBI in terms of sound contrast between the object and the background environment. It was observed that the imaging quality decreased as the sound contrast increased. This can be attributed to the limitations of the Born approximation, which causes divergence when $\Delta\varphi > \pi$. Consequently, the incident frequency (f) must be $< \frac{c_0}{2d \times \% \Delta c}$, where c_0 is the wave propagation speed in the background medium, d is the object diameter, and $\%\Delta c$ is the sound contrast. In the present study, with $c_0 = 1540$ m/s and d = 7.3 mm, and sound contrasts of 10%, 15%, 20%, 25%, and 30%, the required frequencies f were <1.05, 0.70, 0.53, 0.42 and 0.35 MHz, respectively. The simulation set used in the study had a transmitter frequency of 0.64 MHz, which satisfied the Born approximation condition for sound contrasts of 10% and 15%. For these cases, the reconstructed image quality was good, and a small normalized error was obtained. However, for sound contrasts of 20%, 25%, and 30%, where the Born condition was not met, the normalized error increased. Nevertheless, the study addressed the challenging problem of reconstructing images under a high contrast of 30%, where the object of interest was still successfully reconstructed, albeit with some artifacts near its center. Thus, the study demonstrated a viable solution to the Born approximation method, which was also investigated in a previous study (Kak & Slaney, 2001).

Figure 20. Effect of the number of measurements on the quality of the reconstructed image.

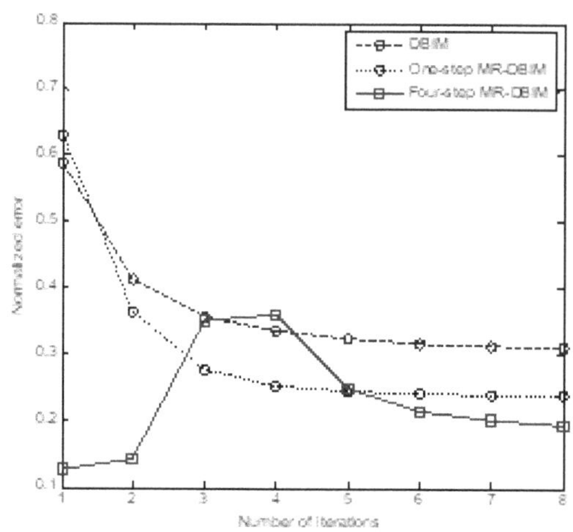

Figure 21. Quality of the reconstructed image using MR-MF-DBI in terms of the sound contrast between the object and background environment.

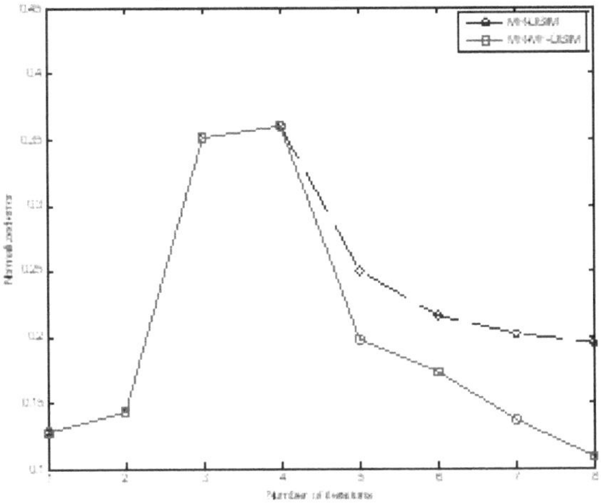

Figure 22 and 23 depict the reconstructions resulting from the application of MR-DBI (Figs. 22a and 23a) and MR-MF-DBI (Figs. 22b and 23b) techniques after the eighth iteration, for the scenarios involving two and three cylinders within the region of interest, respectively. The corresponding Relative Reconstruction Errors (RREs) for MR-DBI and MR-MF-DBI methods were 0.3828 and 0.2880 for the case of two cylinders, and 0.4577 and 0.3579 for the case of three cylinders, respectively. A comparison between the two methods revealed a reduction of 24.76% and 21.80% in RREs for the MR-MF-DBI method, in the case of two and three cylinders, respectively, when compared to MR-DBI. Furthermore, the visual examination of the reconstructed images reveals a significant reduction of noise in both the background and the objects, particularly in the case involving two cylinders.

The observed phenomenon may be attributed to the incremental frequency increment, which has a noise-reducing effect and enhances the resolution of the resulting image. Initially, the original image is recovered with a contrast of c_1 at a lower frequency, while a higher frequency enables the recovery of an image with a contrast of c_2, which surpasses c_1. By continuously augmenting the frequency, it is possible to attain the desired level of contrast in the object of interest, i.e., c_*, where $c_1 < c_2 < \ldots < c_*$. The significance of this gradual escalation of frequency lies in its potential to enhance the resolution of the reconstructed image, which may ultimately lead to image restoration at the level of biological tissue.

*Figure 22. Reconstructions obtained for the two-cylinders case after the eighth iteration using (**a**) MR-DBI and (**b**) MR-MF-DBI.*

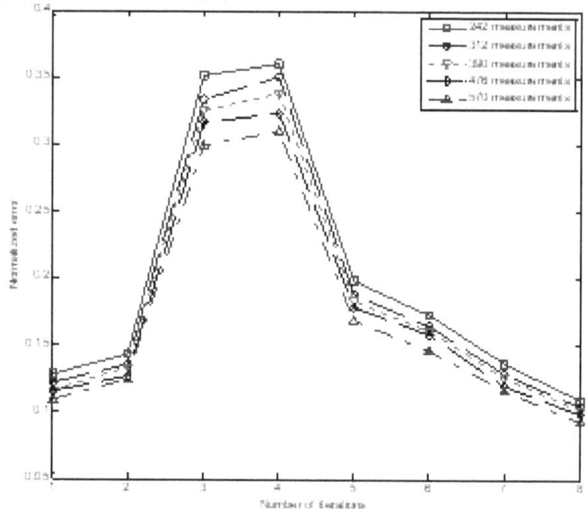

*Figure 23. Reconstructions obtained for the three-cylinders case after the eighth iteration using (**a**) MR-DBI and (**b**) MR-MF-DBI.*

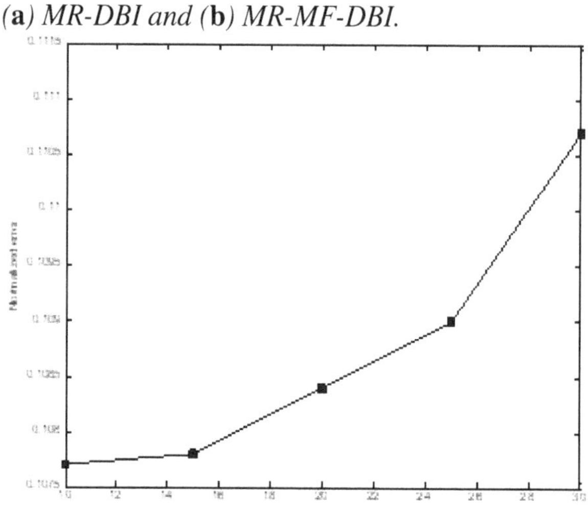

In the field of biomedical imaging, the DBI has gained significant attention for its potential applications in medical diagnostics, particularly in conditions where precise reconstruction of tissue properties is essential. One of the most promising areas where DBI is being explored is in breast cancer detection, where the ability to differentiate between healthy and malignant tissues based on their dielectric properties plays a crucial role in early diagnosis (Costanzo & Flores, 2022). In geophysical exploration, the DBI has been utilized for reconstructing subsurface properties. However, recent advancements have explored the integration of deep learning techniques to enhance seismic imaging. For instance, the work (Zheng & Wang, 2023) introduced a high-resolution reflection seismic imaging method to reveal subsurface geologic structures of a deep geothermal reservoir, demonstrating the potential of reflection seismic imaging in assessing geothermal energy development.

Despite these promising applications, several challenges hinder the practical implementation of DBI. One of the primary limitations is the ill-posed nature of the inverse scattering problem, making DBI susceptible to local minima and instability in reconstructions. To address this issue, researchers often employ regularization techniques, such as Tikhonov regularization and sparsity constraints, to stabilize the inversion process and improve solution accuracy. Another major challenge is computational complexity. DBI requires multiple forward and inverse calculations per iteration, leading to high computational costs, especially for large-scale 3D problems. This limitation makes DBI difficult to implement in real-time applications. However, recent advancements, such as GPU acceleration and parallel computing, have shown promise in reducing computation time and making DBIM more feasible for practical use. Moreover, DBIM is highly sensitive to measurement noise and limited data availability. In real-world scenarios, incomplete sensor coverage and external disturbances can degrade the accuracy of reconstructions. Techniques such as beamforming-enhanced DBI and multi-frequency reconstruction have been proposed to enhance robustness and minimize errors. These approaches allow DBIM to perform better in environments with limited aperture and frequency constraints, improving its practicality for field applications. Finally, the requirement for an accurate initial guess remains a significant challenge in DBI. Since DBI is an iterative method, the quality of the final reconstruction heavily depends on the initial permittivity estimate. Poor initialization can lead to slow convergence or inaccurate results. To overcome this issue, recent studies have explored the use of deep learning-based approaches, such as U-Net convolutional neural networks (CNNs), to generate high-quality initial guesses. By integrating AI-driven initialization techniques, DBI has shown improved accuracy and faster convergence rates, making it more effective in solving complex inverse scattering problems. Despite these challenges, DBI remains a powerful tool for imaging and inverse scattering applications. By addressing issues related to computational efficiency, noise sensitivity, and initialization, DBI can be further optimized for real-world biomedical, geophysical, and security applications. Future research could focus on integrating machine learning, advanced regularization techniques, and adaptive algorithms to improve its robustness and expand its practical usability.

6. CONCLUSION

In this study, we introduce three novel signal processing techniques, namely interpolation, dual-frequency, multi-resolution, and multi-frequency DBI (MR-MF-DBI), for ultrasound tomography. Through simulation, we have demonstrated that these methods significantly improve the performance of ultrasound tomography. However, to apply these methods in real-life scenarios, further research using experimental data is required.

Regarding MR-DBI, its effectiveness is only observed when $r < 1$ (i.e., the scattering domain is sparse). The best performance is achieved when the number of transmitters is equal to the number of receivers in DBI, but in MR-DBI, when the number of receivers is twice the number of transmitters. Moreover, the optimal value of N_{N1} is 2 when $0.5 < r < 1$ and is 3 when $0.2 < r < 0.5$.

We conducted an analysis of the effect of the dual-frequency combination method on the quality of sound contrast-based ultrasound tomography in the context of DF-DBI. We established constraints on simulated scenarios to control computational costs and determined the optimal iteration number N_{f_1} to achieve the best performance. Our findings indicate that the DF-DBI method is only effective for a moderate number of iterations, transmitters, and receivers. When the number of transmitters and receivers is either too large or too small, the DF-DBI approach does not offer superior reconstruction outcomes compared to the single frequency method. Based on these observations, we further investigated more realistic scenarios and found that the best value of N_{f_1}, which is $N_{iter}/2$, reduces the normalized error by 67.6% compared to the conventional single frequency approach.

This study introduces the multi-resolution technique in the context of MR-MF-DBI, which employs a four-step interpolation scheme to accelerate and improve the imaging quality of the object of interest. The results of our imaging analysis showed that the relative root mean square error (RRE) was reduced by 18.67% and 37.21% compared to the multi-resolution technique (one-step interpolation) and DBI methods, respectively. Moreover, we explored the integration of multi-resolution and multi-frequency techniques for high-contrast object imaging, and found that the normalized absolute error was reduced by 44.01% compared to the MR-DBI method.

REFERENCES

Abubakar, A., Habashy, T. M., van den Berg, P. M., & Gisolf, D. (2005). The diagonalized contrast source approach: An inversion method beyond the Born approximation. *Inverse Problems*, *21*(2), 685–702. DOI: 10.1088/0266-5611/21/2/015

Avinash, C. K., & Slaney, M. (2001). *Principles of computerized tomographic imaging*. IEEE Press-Society for Industrial and Applied Mathematics.

Born, M., & Wolf, E. (2013). *Principles of optics: Electromagnetic theory of propagation, interference and diffraction of light* (7th ed.). Elsevier.

Costanzo, S., & Flores, A. (2022). Enhanced machine learning approach for accurate and fast resolution of inverse scattering problem in breast cancer detection. *Electronics (Basel)*, *11*(15), 2308. DOI: 10.3390/electronics11152308

Devaney, A. J. (1982). Inversion formula for inverse scattering within the Born approximation. *Optics Letters*, *7*(3), 111–112. DOI: 10.1364/OL.7.000111 PMID: 19710840

Ghai, D., Tripathi, S. L., Saxena, S., Chanda, M., & Alazab, M. (2022). *Machine learning algorithms for signal and image processing*. Wiley IEEE Press. DOI: 10.1002/9781119861850

Golub, G. H., Hansen, P. C., & O'Leary, D. P. (1999). Tikhonov regularization and total least squares. *SIAM Journal on Matrix Analysis and Applications*, *21*(1), 185–194. DOI: 10.1137/S0895479897326432

Greenleaf, J. F., & Bahn, R. C. (1981). Clinical imaging with transmissive ultrasonic computerized tomography. *IEEE Transactions on Biomedical Engineering*, *2*(2), 177–185. DOI: 10.1109/TBME.1981.324789 PMID: 7287021

Haddadin, O. S., & Ebbini, E. S. (1998). Imaging strongly scattering media using a multiple frequency distorted Born iterative method. *IEEE Transactions on Ultrasonics, Ferroelectrics, and Frequency Control*, *45*(6), 1485–1496. DOI: 10.1109/58.738288 PMID: 18249996

Haddadin, O. S., & Emad, S. E. (1997). Multiple frequency distorted Born iterative method for tomographic imaging. *Acoustical Imaging*, *23*, 613–619. DOI: 10.1007/978-1-4419-8588-0_96

Haddadin, O. S., & Emad, S. E. (1998). Imaging strongly scattering media using a multiple frequency distorted Born iterative method. *IEEE Transactions on Ultrasonics, Ferroelectrics, and Frequency Control, 45*(6), 1485–1496. DOI: 10.1109/58.738288 PMID: 18249996

Haddadin, O. S., Lucas, S. D., & Ebbini, E. S. (1995). Solution to the inverse scattering problem using a modified distorted Born iterative algorithm. *In1995 IEEE Ultrasonics Symposium Proceedings: An International Symposium* (pp. 1411-1414). IEEE. DOI: 10.1109/ULTSYM.1995.495819

Hesford, A. J., & Chew, W. C. (2010). Fast inverse scattering solutions using the distorted Born iterative method and the multilevel fast multipole algorithm. *The Journal of the Acoustical Society of America, 128*(2), 679–690. DOI: 10.1121/1.3458856 PMID: 20707438

Highnam, R. P., Brady, J. M., & Shepstone, B. J. (1995). A representation for mammographic image processing. *InInternational Conference on Computer Vision, Virtual Reality, and Robotics in Medicine* (pp. 365-371). Springer, Berlin, Heidelberg.

Huy, T. Q.. (2019). The efficiency of applying compressed sampling and multiresolution into ultrasound tomography. *Ingeniería Solidaria, 15*(3), 1–16.

Huy, T. Q., Tan, T. D., & Linh-Trung, N. (2014). An improved distorted Born iterative method for reduced computational complexity and enhanced image reconstruction in ultrasound tomography. In *2014 International Conference on Advanced Technologies for Communications (ATC 2014)* (pp. 703-707). IEEE.

Kak, A., & Slaney, M. (2001). Principles of computerized tomographic imaging. SIAM: University City, PA, USA. DOI: 10.1137/1.9780898719277

Krainov, V. P., Reiss, R. H., & Smirnov, M. B. (2005). Appendix J: Properties of the generalized Bessel function. In Radiative Processes in Atomic Physics (pp. 205-210). John Wiley & Sons.

Lavarello, R., & Oelze, M. (2009). Tomographic reconstruction of three-dimensional volumes using the distorted Born iterative method. *IEEE Transactions on Medical Imaging, 28*(1), 1643–1653. DOI: 10.1109/TMI.2009.2026274 PMID: 19574162

Madsen, E. L., Zagzebski, J. A., Frank, G. R., Greenleaf, J. F., & Carson, P. L. (1982). Anthropomorphic breast phantoms for assessing ultrasonic imaging system performance and for training ultrasonographers: Part I. *Journal of Clinical Ultrasound, 10*(2), 67–75. DOI: 10.1002/jcu.1870100207 PMID: 6804504

Magdum, A. D., Shimoga Beerappa, H., & Erramshetty, M. (2024). Deep learning based distorted Born iterative method for improving microwave imaging. *Frequenz (Bern)*, *78*(1-2), 1–8. DOI: 10.1515/freq-2023-0074

Mojabi, P., & LoVetri, J. (2015). Ultrasound tomography for simultaneous reconstruction of acoustic density, attenuation, and compressibility profiles. *The Journal of the Acoustical Society of America*, *134*(4), 1813–1825. DOI: 10.1121/1.4913774 PMID: 25920834

Slaney, M., Kak, A. C., & Larsen, L. E. (1984). Limitations of imaging with first-order diffraction tomography. *IEEE Transactions on Microwave Theory and Techniques*, *32*(8), 860–874. DOI: 10.1109/TMTT.1984.1132783

Tijhuis, A. G., Belkebir, K., Litman, A. C., & de Hon, B. P. (2001). Multiple-frequency distorted-wave Born approach to 2D inverse profiling. *Inverse Problems*, *17*(6), 1635–1644. DOI: 10.1088/0266-5611/17/6/307

Tijhuis, A. G., Belkebir, K., Litman, A. C., & Hon, B. P. (2001). Multiple-frequency distorted-wave Born approach to 2D inverse profiling. *Inverse Problems*, *17*(6), 1635–1644. DOI: 10.1088/0266-5611/17/6/307

Tran-Duc, T., Nguyen-Tien, A., & Anh, T. (2011). Application of interpolation for DBI reconstruction of ultrasound tomography. In *2011 International Symposium on Micro-NanoMechatronics and Human Science* (pp. 240-243). IEEE.

Wiskin, J., Borup, D. T., Johnson, S. A., & Berggren, M. (2012). Non-linear inverse scattering: High-resolution quantitative breast tissue tomography. *The Journal of the Acoustical Society of America*, *131*(5), 3802–3813. DOI: 10.1121/1.3699240 PMID: 22559356

Xing, Z., Zhang, Z., Chen, Z., Wang, Y., Ma, H., & Wei, Z. (2025). VBIM-Net: Variational Born Iterative Network for inverse scattering problems. *IEEE Transactions on Geoscience and Remote Sensing*, *63*, 1–16. DOI: 10.1109/TGRS.2025.3532363

Yen, N. T. H. (2018). An improvement for tomographic density imaging using integration of DBI and interpolation. In *2018 International Conference on Research in Intelligent and Computing in Engineering (RICE)* (pp. 1-4). IEEE.

Zheng, Y., & Wang, Y. (2023). High-resolution reflection seismic imaging to reveal subsurface geologic structures of a deep geothermal reservoir. *Geophysics*, *88*(5), WB37–WB43. DOI: 10.1190/geo2022-0494.1

Ziegler, S., & Burfeindt, M. (2025). Frequency beamforming-enhanced DBIM for limited-aperture quantitative imaging. *IEEE Transactions on Antennas and Propagation*, *73*(2), 1173–1184. DOI: 10.1109/TAP.2025.3526904

Chapter 4

Examining Personal Lifestyle Factors and Lipid Profile on Cardiovascular Disease Risk Prediction

Pawan Kumar Badhan
https://orcid.org/0009-0002-3953-6825
Pyramid College of Business and Technology, Phagwara, India

ABSTRACT

Cardiovascular disease (CVD) remains a significant global health concern, influenced by factors like lifestyle, lipid profiles, and genetics. Early prediction is vital for effective prevention and management. This study develops predictive models for CVD risk by combining demographic, health, genetic, and imaging data, utilizing machine learning (ML) and deep learning (DL) techniques. The main objective is to enhance risk prediction accuracy by analyzing X-ray image data through convolutional neural networks (CNNs). Additionally, the study explores the integration of structured patient data (demographics, lipid profiles, genetics) with imaging features to provide a comprehensive CVD risk assessment. The dataset, consisting of 264 patients, includes diverse factors like lifestyle and health indicators. Data preprocessing involved normalization, resizing, and feature selection through techniques like LASSO. ML models (SVM, RF, DT) achieved up to 95% accuracy, while CNNs analyzing X-ray images reached 100% accuracy.

DOI: 10.4018/979-8-3373-0548-6.ch004

1. INTRODUCTION

Cardiovascular disease (CVD) remains a leading cause of morbidity and mortality worldwide, necessitating advanced predictive models for early detection and prevention (Dalal et al., 2023) (Peng et al., 2023). Conventional pitfall assessment procedures essentially utilize clinical evidence measures and focus on basic characteristics, often overlooking the intricate interplay of personal lifestyle factors and lipid profiles (Ordikhani et al., 2022) (Dritsas et al., 2022). In recent years, machine learning (ML) and deep learning (DL) have emerged as powerful tools in medical research, offering the potential to enhance CVD risk prediction by integrating complex, multidimensional data (Qian et al., 2022) (Chinnasamy et al., 2022). ML and DL algorithms are adept at managing large datasets, discerning subtle patterns, and delivering precise predictions (Pal et al., 2022) (Ansarullah et al., 2022). These technologies can integrate diverse data sources, such as dietary habits, physical activity levels, smoking status, and detailed lipid profiles, to offer a more comprehensive risk assessment (Ramesh et al., 2022) (Badhan, 2024). By harnessing the predictive power of ML and DL, researchers can reveal hidden connections between lifestyle behaviours, lipid levels, and CVD risk, thereby enabling personalized intervention strategies (Swathy & Saruladha, 2022) (Uddin & Halder, 2021). This study highlights the potential of machine learning models in refining risk assessment for CVD incidence, underlining the importance of precise predictive algorithms for bolstering disease control and management strategies (Hossen et al., 2021) (Rahim et al., 2021).

This study aims to explore the application of ML and DL in predicting CVD risk by integrating personal lifestyle factors and lipid profiles (Kim et al., 2021) (Faizal et al., 2021). The objective is to develop models that not only improve predictive accuracy but also provide actionable insights into modifiable risk factors, thereby enhancing preventive healthcare measures and reducing the global burden of cardiovascular diseases (Islam et al., 2020) (Krittanawong et al., 2020). This research endeavours to delve into the nuanced relationship between personal lifestyle factors and CVD risk using ML algorithms as the analytical framework (Doust et al., 2020) (Quesada et al., 2019). By using a large dataset that includes information on demographics, health, genes, and habits, this study aims to elucidate how these factors collectively contribute to the prediction of CVD risk and increase accuracy in predicting desired results (Dinh et al., 2019) (Pate et al., 2019). The study employs a diverse range of ML algorithms, including DT, RF, and CNNs, to comprehensively analyse the dataset (Badhan & Kaur, 2024) (Meshref, 2019). Each algorithm undergoes rigorous training and evaluation to assess its predictive accuracy and performance in forecasting CVD risk (Alaa et al., 2019) (Chandralekha & Shenbagavadivu, 2018). Preliminary findings suggest promising results, with several ML algorithms exhibiting high levels of accuracy in predicting CVD risk (Dimopoulos et al., 2018) (Chiuve et al., 2014). Decision trees and support vector machines have demonstrated perfect accuracy, while random forests and neural networks have achieved notably high accuracy rates (Qian et al., 2022), (Pal et al., 2022). However, convolutional neural networks (CNNs) show a slightly lower accuracy, indicating potential areas for improvement (Ramesh et al., 2022), (Badhan & Kaur, 2024).

Future research could delve deeper into this potential by investigating additional variables, refining algorithms, and developing personalized risk assessment models (Dalal et al., 2023), (Krittanawong et al., 2020). These efforts offer the potential to improve patient outcomes and reduce the global impact of cardiovascular diseases by implementing tailored preventive interventions (Quesada et al., 2019), (Alaa et al., 2019). In addition to exploring the role of personal lifestyle factors, understanding the influence of lipid profiles on cardiovascular disease (CVD) risk prediction is crucial (Ansarullah et al., 2022), (Kim et al., 2021). Lipid profiles, encompassing cholesterol levels and other lipid biomarkers, serve as significant indicators of cardiovascular health (Uddin & Halder, 2021), (Dimopoulos et al., 2018). High levels of low-density lipoprotein (LDL) cholesterol, for instance, are associated with increased risk of atherosclerosis and coronary artery disease, while elevated levels of high-density lipoprotein (HDL) cholesterol are often considered protective against CVD (Ordikhani et al., 2022), (Qian et al., 2022). Moreover, personal lifestyle factors such as diet, physical activity, smoking habits, and stress levels profoundly impact lipid profiles and overall cardiovascular health (Chinnasamy et al., 2022). Unhealthy dietary patterns rich in saturated fats and processed sugars can elevate LDL cholesterol levels, while regular exercise and a balanced diet can help maintain optimal lipid profiles (Ansarullah et al., 2022; Doust et al., 2020). Integrating information from both personal lifestyle factors and lipid profiles into predictive models can significantly enhance CVD risk prediction accuracy (Hossen et al., 2021; Pate et al., 2019). Machine learning and deep learning techniques offer a promising approach to analyse the complex interactions between these variables and their collective impact on cardiovascular health (Faizal et al., 2021; Chiuve et al., 2014). By incorporating comprehensive datasets that include lipid profiles alongside lifestyle factors, researchers can develop more nuanced predictive models capable of identifying individuals at heightened risk of developing CVD (Alaa et al., 2019; Chandralekha & Shenbagavadivu, 2018). This holistic approach to risk assessment facilitates targeted interventions aimed at promoting healthier lifestyles, managing lipid profiles, and ultimately reducing the incidence and severity of cardiovascular diseases (Kim et al., 2021; Meshref, 2019).

Moreover, the architecture in Figure 1 presents a comprehensive framework for predicting cardiovascular disease (CVD) risk by integrating personal lifestyle factors and lipid profiles using machine learning (ML) and deep learning (DL) methods.

The process starts with Input Data, which includes a variety of datasets such as demographic details (e.g., age, gender), health records (e.g., medical history), lifestyle factors (e.g., exercise habits, diet, smoking status), lipid profiles (e.g., cholesterol levels), and X-ray images. These variables collectively provide a holistic view of an individual's cardiovascular health, highlighting potential risk factors.

Figure 1. Integrating Personal Lifestyle Factors and Lipid Profiles for Cardiovascular Disease Risk Prediction Using Machine Learning and Deep Learning.

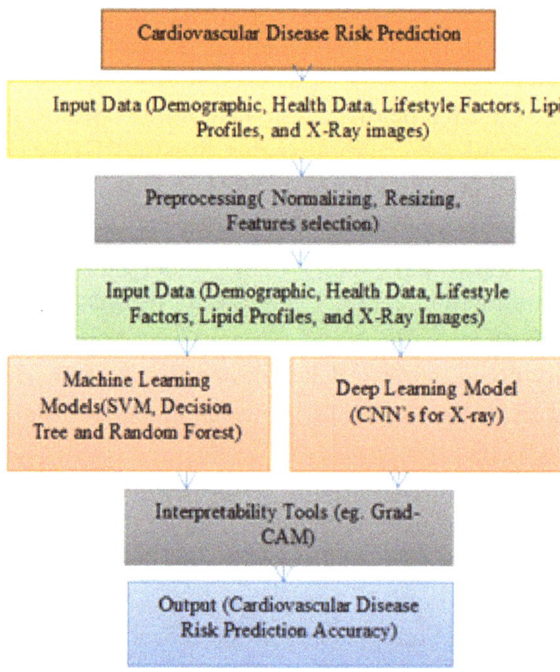

The data then proceeds to the Preprocessing stage, where normalization is applied to ensure that all features are scaled uniformly, preventing dominance by any single variable due to differences in units or scale. Additionally, resizing is performed on X-ray images to standardize their dimensions, and feature selection helps identify the most significant predictors of CVD risk, eliminating irrelevant or redundant data. Following preprocessing, the data is divided into two parallel paths: Machine Learning Models and Deep Learning Models. In the Machine Learning Models branch, algorithms such as DT, RF, and Support SVM are employed to predict CVD risk based on structured, tabular data. These models are valued for their interpretability and efficiency. In contrast, the Deep Learning Models path utilizes CNNs to analyze X-ray images, enabling the extraction of intricate patterns that may not be captured by traditional methods. To improve transparency, Interpretability Tools like Grad-CAM are used, which visually highlight key features of the X-ray images, aiding in understanding model predictions.

In the context of this study, feature selection techniques are utilized to pinpoint the most predictive variables from the comprehensive dataset encompassing personal lifestyle factors and lipid profiles (Balamurugan et al., 2021; Rajagopal et al., 2022). Techniques such as statistical tests, correlation analysis, and machine learning-based feature importance algorithms are utilized to prioritize the most relevant predictors for inclusion in the predictive models (Al Bashish et al., 2021; Hossen et al., 2021). Furthermore, the inclusion of lipid profile data in this analysis represents a novel and essential aspect of this research chapter (Alzubaidi et al., 2021; Kumar & Choudhury, 2020). While previous studies have predominantly focused on lifestyle factors alone, integrating lipid profiles provides a more comprehensive understanding of cardiovascular risk factors (Iwendi et al., 2020; Chiuve et al., 2014). By incorporating lipid biomarkers such as cholesterol levels and triglycerides into the predictive models, the research aims to provide accurate and robust predictions of CVD risk (Kaur & Kumari, 2021; Islam et al., 2021). Looking ahead, the incorporation of lipid profile data opens up exciting avenues for future research (Doust et al., 2020; Sarwar et al., 2017). One potential direction is to explore advanced machine learning techniques, such as deep learning architectures, specifically tailored for analyzing lipidomic data (Ramalingam et al., 2021; Suresh et al., 2020). Additionally, longitudinal studies could investigate the dynamic changes in lipid profiles over time and their implications for long-term cardiovascular health (Abdar et al., 2019; Balamurugan et al., 2021).

Literature Review

Cardiovascular diseases (CVD) remain a significant public health concern globally, necessitating effective risk assessment and prediction strategies. In recent years, researchers have increasingly turned to feature selection techniques as a means to minimize prediction errors and effectiveness of predictive models for CVD risk. In the research article by Dalal and Goel et al. (2023), the application of machine learning for cardiovascular disease risk prediction is explored. The study utilized a dataset from Kaggle containing approximately 70,000 patient records with 11 features to forecast cardiovascular disease. Various machine learning models, including neural networks, random forests, Bayesian networks, C5.0, and QUEST, were compared, achieving a high accuracy of 99.1 percent. Peng and Hou et al. (2023), the authors developed the XGBH machine learning model for predicting cardiovascular disease (CVD) risk using key contributing features. By combining retrospective data of 14,832 Chinese Shanxi CVD patients with the Kaggle dataset, the model achieved a high accuracy (AUC = 0.81), outperforming the baseline risk score (AUC = 0.65). Additionally, a simpler model using only age, systolic blood pressure, and cholesterol status maintained a high accuracy (AUC = 0.79). Study by Ordikhani and Abadeh et al. (2022), the authors developed a novel risk assessment model for predicting cardiovascular disease (CVD) events using a Genetic Algorithm (GA). The model was calibrated using the Isfahan Cohort Study (ICS) database, which included 6,504 healthy Iranian adults followed over ten years. The GA-based model, named eXplanaible Persian Atherosclerotic CVD Risk Stratification (XPARS), outperformed traditional statistical methods and other machine learning models. The XPARS model with eight features achieved an AUROC of 0.76, while a simplified version with four features achieved an AUROC of 0.72. Dritsas and Alexiou et al. (2023), the authors evaluated supervised machine learning techniques for predicting cardiovascular disease (CVD) risk in participants over 50. The Logistic Regression model outperformed Naive Bayes, SVM, and Random Forest, achieving 72.1% accuracy and an AUC score of 78.4%, making it the most effective for early CVD prediction. Qian and Li et al. (2023), a cardiovascular disease (CVD) prediction model was developed using machine learning algorithms based on routine physical examination indicators for the Xinjiang rural population. The cohort included 12,692 participants, and after 4.94 years of follow-up, 1,176 cases of CVD were diagnosed (9.27% cumulative incidence). The model based on L1 regularized logistic regression (L1-LR) showed the best prediction performance, with key predictors including age, systolic blood pressure, lipid levels, and body mass index. In the study by Chinnasamy and Kumar et al. (2022), a machine learning approach for predicting cardiovascular disease (CVD) is proposed to improve early detection and risk assessment. The method aims to identify heart abnormalities quickly, facilitating timely interventions and better clinical decision-making. Pal and Parija et al. (2022), machine learning classifiers, specifically multi-layer perceptron (MLP) and K-nearest neighbour (K-NN), were used for cardiovascular disease (CVD) risk prediction. The MLP model achieved an accuracy of 82.47% and an area under the curve (AUC) of 86.41%, outperforming K-NN. The authors recommend the MLP model for automatic CVD detection. Ansarullah et al. (2022), a heart disease risk evaluation model using non-invasive risk factors was developed. Various machine learning techniques were tested, with the random forest model showing the highest predictive accuracy and lowest misclassification rate for heart disease risk assessment. In the study by Ramesh et al. (2022), the authors analyse heart disease prediction using various machine learning methods on a UCI dataset. Focusing on 14 key features, they evaluate Naive Bayes, SVM, logistic regression, decision tree, random forest, and K-nearest neighbour (KNN). The results show that KNN, with eight neighbours, achieves the highest effectiveness, sensitivity, precision, accuracy, and F1-score compared to the other methods. Swathy and Saruladha (2022), various machine learning and deep learning techniques for predicting cardiovascular diseases (CVD) are compared. The paper reviews models, datasets, and performance metrics, highlighting their potential to improve early detection and reduce mortality associated with risk factors like obesity and hypertension. Uddin and Halder et al. (2021) propose an ensemble method-based multilayer dynamic system for predicting cardiovascular disease (CVD) using machine learning. The study emphasizes the effectiveness of integrating multiple models to enhance predictive accuracy and address challenges in early CVD detection. By analyzing various datasets and performance metrics, the authors highlight the potential of their approach in identifying risk factors and improving management strategies for cardiovascular diseases. Hossen and Tazin et al. (2021) examine supervised machine learning techniques for cardiovascular disease analysis and prediction, focusing on various algorithms and their effectiveness. The authors provide insights into model performance, contributing to the understanding of how these methods can enhance predictive capabilities in CVD risk assessment. Rahim and Rasheed et al. (2021) present an integrated machine learning framework designed to improve the prediction of cardiovascular diseases.

Kim and Jeong et al. (2021) develop a machine learning-based prediction model for cardiovascular diseases using data from the Korean national health insurance service health screening database. The study demonstrates the potential of machine learning to identify high-risk individuals, contributing to preventive healthcare strategies for CVD. Faizal and Thevarajah et al. (2021) review risk prediction models for cardiovascular disease, comparing conventional methods with artificial intelligence approaches. The authors discuss the advantages of integrating AI in risk assessment, emphasizing its potential to improve prediction accuracy and clinical decision-making. Islam and Jahan et al. (2020) explore various machine learning paradigms for forecasting cardiovascular disease. The paper discusses the applicability of different algorithms and highlights the challenges in data collection and model training, underlining the need for robust methodologies in CVD prediction. Krittanawong and Virk et al. (2020). conduct a meta-analysis on machine learning predictions in cardiovascular diseases, synthesizing findings from multiple studies. The authors assess the overall performance of machine learning models, concluding that they offer promising results in risk prediction compared to traditional methods. Doust and Bonner et al. (2020) discuss future directions in cardiovascular disease risk prediction, emphasizing the need for innovative approaches to enhance accuracy and applicability in clinical settings. The authors advocate for the integration of advanced technologies, including machine learning, to improve patient outcomes. Quesada and Lopez-Pineda et al. (2019) investigate the use of machine learning to predict cardiovascular risk, highlighting various algorithms and their effectiveness in risk stratification. The study suggests that machine learning can enhance traditional risk assessment methods, providing valuable insights for preventive healthcare. Dinh and Miertschin et al. (2019) adopt a data-driven approach to predict diabetes and cardiovascular disease using machine learning. The authors analyze different algorithms and emphasize the importance of data quality and feature selection in achieving reliable predictions. Pate and Emsley et al. (2019) address the uncertainties associated with risk prediction models for individual decision-making in cardiovascular disease. Through an exemplar cohort study, the authors highlight the challenges in translating model outputs into clinical practice, emphasizing the need for clearer guidelines. Meshref et al. (2019) proposes a machine learning interpretation approach for cardiovascular disease diagnosis. The study emphasizes the importance of understanding model predictions, contributing to the development of more transparent and interpretable AI applications in healthcare. Alaa and Bolton et al. (2019) focus on cardiovascular disease risk prediction using automated machine learning, analyzing data from over 423,000 UK Biobank participants. The study illustrates the potential of automated approaches to enhance risk stratification and inform clinical decision-making. Chandralekha and Shenbagavadivu et al. (2018) provide an empirical study comparing various machine learning techniques for predicting cardiovascular disease. The authors evaluate model performance across different datasets, contributing valuable insights into the strengths and weaknesses of each approach. Dimopoulos and Nikolaidou et al. (2018) compare machine learning methodologies with traditional cardiovascular risk scores in predicting disease risk. The study highlights the advantages of machine learning in providing more accurate predictions and encourages further research in this area. Chiuve and Cook et al. (2014) develop a lifestyle-based prediction model for cardiovascular disease prevention, known as the Healthy Heart Score. The authors demonstrate how lifestyle factors can significantly influence risk assessment and advocate for the integration of lifestyle modifications in preventive strategies.

Through an extensive review, we've gathered insights to refine cardiovascular disease (CVD) identification using lipid profiles and lifestyle factors. In this, the study intend to utilize a new dataset consisting of information gathered from 264 patients obtained from clinics and hospitals. By utilizing this dataset, research anticipate enhancing CVD identification accuracy through machine learning algorithms and convolutional neural networks (CNNs). This dataset offers several advantages, including a larger patient cohort for robust analysis, diverse data sources for comprehensive patient profiles, and inclusion of lipid profiles and lifestyle factors for holistic risk assessment. Employing decision tree, support vector machines, random forest, and CNN algorithms will enable accurate risk prediction based on multidimensional data.

2. MATERIAL AND METHODS SELECTION

Cardiovascular diseases (CVD) are one of the primary causes of mortality worldwide, emphasizing the urgent need for effective risk assessment and prediction strategies. Personal lifestyle factors, including diet, physical activity, and smoking habits, significantly influence the onset and progression of CVD. As such, it is crucial to integrate these factors into predictive models to better identify individuals at heightened risk. To achieve this objective, this study will utilize a detailed dataset that encompasses demographic information, health indicators, genetic markers, and lifestyle factors. This dataset will allow us to evaluate how well different machine learning algorithms can predict CVD risk, taking into account the interplay between personal lifestyle choices and lipid profiles. By employing a rigorous selection of machine learning techniques—including Decision Trees (DT), Random Forests (RF), and Convolutional Neural Networks (CNNs)— seek to demonstrate the potential of these algorithms in accurately forecasting the likelihood of developing cardiovascular diseases. This research will detail the materials and methods employed, including a thorough description of the dataset composition, the machine learning algorithms utilized, and the evaluation metrics applied to measure predictive performance. This comprehensive methodology aims to provide valuable insights into how personal lifestyle factors influence CVD risk assessment and to demonstrate the effectiveness of machine learning techniques in improving predictive accuracy in this domain.

2.1 Dataset

This dataset includes a wide range of variables essential for analyzing how personal lifestyle factors and lipid profiles impact cardiovascular disease (CVD) risk prediction of 264 patients from number of clinical trials from different health care homes in Table 1. The data collected for predicting CVD risk through ML models encompasses a panoramic range of criteria that employs on various methods to ensure thoroughness and relevance. Each entry in the dataset corresponds to an individual, identified by a unique ID number. Basic demographic details such as age and gender are included, with gender coded as 1 for women and 2 for men. Physical characteristics are recorded through height (in centimetres) and weight (in kilograms), which are used to calculate the body mass index (BMI). Blood pressure readings include systolic (ap_hi) and diastolic (ap_lo) pressures. Cholesterol and glucose levels are classified into three categories: normal, above normal, and well above normal. Smoking status is indicated by binary values (0 for no and 1 for yes).

Lifestyle factors are detailed, covering dietary habits (e.g., balanced diet, low in saturated fats, high consumption of processed foods, Mediterranean diet), physical activity levels (ranging from sedentary to very active), and alcohol consumption (none, moderate, heavy). It also notes the use of medications related to lipid or cardiovascular health (e.g., statins, blood pressure medication) and dietary supplements (e.g., fish oil, fiber supplements, multivitamins). Measurements of waist circumference (in centimetres), physical exam findings (normal or abnormal), mental health status (no mental health issues, anxiety, depression, stress), and sleep quality (poor, average, good) are included. Socioeconomic status is classified as low, middle, or high.

The dataset in Table 1 also records detailed dietary patterns (vegetarian, vegan, paleo, keto, high carb, low carb) and ethnic background (specific categories such as Caucasian, African American, Asian, Hispanic). Blood lipid levels are detailed with numeric values for LDL, HDL, and triglycerides. Resting heart rate (in beats per minute) is provided. It includes history of previous cardiovascular events (none, heart attack, stroke, other) and levels of inflammatory markers (e.g., CRP, ESR). Occupational risk factors (none, exposure to stress, sedentary job, exposure to pollutants) and environmental factors (urban, suburban, rural) are also included.

Table 1. Integrated Dataset for Cardiovascular Disease Risk Prediction: Incorporating Lipid Profiles and Personal Lifestyle Factors.

Fields	Description
Patient_Age	Age of the individual in years.
Patient Gender	Gender of the individual. Values: 1: Women, 2: Men
Patient Height	Height of the individual in cm.
Patient_Weight	Weight of the individual in kg.
ap_hi	Systolic blood pressure.
ap_lo	Diastolic blood pressure.
Cholesterol	Cholesterol level. Values: 1: Normal, 2: Above normal, 3: Well above normal.
Gluc	Glucose level. Values: 1: Normal, 2: Above normal, 3: Well above normal
Smoking	Indicates whether the patient smokes. Values: 0: No, 1: Yes
Cardiovascular Risk Score	Calculated cardiovascular risk score based on lipid profile and other risk factors. Values: Moderate, High, Low
Family History	Family history of cardiovascular diseases related to lipid disorders. Values: Yes, No
Lab Tests	Results of additional laboratory tests relevant to lipid profile. Values: Fasting glucose: 100, Liver function tests normal, Normal lipid panel
Genetic Markers	Genetic variants associated with lipid metabolism or cardiovascular risk. Values: rs7412(T; T), rs1800562(A; G), rs429358(C; T)
Comorbidities	Presence of other medical conditions affecting lipid metabolism or cardiovascular health. Values: Hypertension, Type 2 Diabetes, Obesity, Hyperlipidemia, None
Dietary Habits	Dietary intake relevant to lipid metabolism. Values: Balanced diet, Low in saturated fats, High consumption of processed foods, Mediterranean diet.

continued on following page

Table 1. Continued

Fields	Description
Physical Activity Level	The individual's level of physical activity. Values: Sedentary, Lightly active, Moderately active, Very active.
Alcohol Consumption	Indicates the individual's alcohol consumption. Values: None, Moderate, Heavy.
Medication	Medications taken by the individual relevant to lipid or cardiovascular health. Values: Statins, Blood pressure medication, None.
Dietary Supplements	Use of dietary supplements relevant to cardiovascular health. Values: Fish oil, Fiber supplements, Multivitamins, None
Body Mass Index (BMI)	Calculated Body Mass Index. Values: Numeric value calculated from height and weight.
Waist Circumference	Waist circumference in cm.
Physical Exam Findings	Results of physical examinations relevant to cardiovascular health. Values: Normal, Abnormal findings (e.g., heart murmurs, edema).
Mental Health Status	Information on mental health, which can impact cardiovascular health. Values: No mental health issues, Anxiety, Depression, Stress
Sleep Quality	Quality and duration of sleep. Values: Poor, Average, Good
Socioeconomic Status	Socioeconomic factors that might influence health. Values: Low, Middle, High.
Dietary Patterns	More detailed breakdown of dietary patterns. Values: Vegetarian, Vegan, Paleo, Keto, High carb, Low carb.
Ethnicity	Ethnic background of the individual. Values: Specific ethnic categories (e.g., Caucasian, African American, Asian, Hispanic, etc.).
Blood Lipid Levels	Detailed lipid profile including LDL, HDL, and triglycerides. Values: Numeric values for LDL, HDL, Triglycerides.
Heart Rate	Resting heart rate in beats per minute. Values: Numeric value
Previous Cardiovascular Events	History of previous cardiovascular events like heart attack or stroke. Values: None, Heart attack, Stroke, Other.
Inflammatory Markers	Levels of inflammatory markers relevant to cardiovascular health. Values: CRP (C-reactive protein) levels, ESR (erythrocyte sedimentation rate).
Occupational Risk Factors	Occupational hazards that might impact cardiovascular health. Values: None, Exposure to stress, Sedentary job, Exposure to pollutants.
Environmental Factors	Environmental influences on health. Values: Urban, Suburban, Rural.

The detailed and multi-dimensional nature of this dataset makes it highly valuable for using machine learning algorithms to predict cardiovascular disease risk. The inclusion of a broad range of variables—from basic demographic data to detailed genetic markers and lifestyle factors—enables comprehensive modelling of CVD risk. Machine learning algorithms can exploit the high-dimensional data to identify complex patterns and interactions between various risk factors that may not be detectable through traditional statistical methods. The dataset's diverse variables, such as genetic markers, lifestyle factors, comorbidities, and detailed lipid profiles, ensure that all critical aspects of cardiovascular health are considered, allowing for more accurate and personalized risk predictions. Detailed categorical and numerical data on cholesterol, glucose levels, and inflammatory markers enable machine learning models to capture subtle variations in risk profiles. Factors like mental health status, sleep quality, socioeconomic status, and environmental influences provide a holistic view of the individual's health, leading to more nuanced predictions that take social and psychological factors into account.

2.2 Methods Selection

To effectively select features for forecast CVD risk using ML algorithms, this research leverages a comprehensive dataset that includes a wide array of variables such as demographic data, health indicators, genetic markers, and lifestyle factors. This rich dataset enables a thorough analysis and robust predictive modelling. Several advanced feature selection methods are employed in Table 2 to ensure that the most relevant and impactful features are used, thereby enhancing the accuracy and efficiency of the predictive models.

One key method is correlation analysis, which helps in identifying relationships between variables. By examining the correlation matrix, researchers can pinpoint which features are strongly correlated with the target variable (CVD risk) and with each other. This step is crucial for understanding the interdependencies among variables and for detecting multicollinearity, which can adversely affect the performance of some machine learning algorithms.

Table 2. Optimizing Cardiovascular Disease Risk Prediction through Advanced Feature Selection Techniques.

Method	Description
Correlation Analysis	Identifies relationships between variables by examining the correlation matrix to pinpoint strong correlations with the target variable and among features.
Variance Thresholding	Removes features with low variance, which are less likely to be useful for prediction.
Recursive Feature Elimination (RFE)	Iteratively selects the most significant predictors by fitting a model and removing the least important features until the desired number is reached.
LASSO (Least Absolute Shrinkage and Selection Operator)	Adds a penalty to the magnitude of coefficients, effectively performing variable selection and regularization.

Another key method is variance thresholding, which removes features with low variance, as such features are less likely to be useful for prediction. Setting a variance threshold helps filter out these less informative features, simplifying the model and potentially improving its performance. Recursive Feature Elimination (RFE) is another technique used to iteratively select the most significant predictors. RFE fits a model and removes the least important feature (or features) until the desired number of features is reached. This method is particularly effective with models that provide a coefficient or feature importance score, such as linear models or tree-based models.

Additionally, LASSO (Least Absolute Shrinkage and Selection Operator) is utilized for feature selection. LASSO adds a penalty equal to the absolute value of the magnitude of coefficients, shrinking some coefficients to zero, effectively performing variable selection and regularization. This enhances the prediction accuracy and interpretability of the resulting model. Using these sophisticated feature selection techniques ensures that predictive models focus on the most relevant features, such as age, gender, BMI, blood pressure readings, cholesterol and glucose levels, smoking habits, family medical history, dietary patterns, physical activity levels, and genetic variations. These selected features capture the critical aspects of cardiovascular health and lifestyle, providing a comprehensive basis for accurate risk prediction. These models can capture complex interactions and dependencies among diverse predictors, significantly enhancing the ability to forecast CVD risk.

2.3 Pre-processing and Taining

The preprocessing for the study involved normalizing, resizing, and augmenting datasets to enhance model robustness. Demographic data, health indicators, genetic markers, and personal lifestyle factors in Tvale 3 were meticulously curated from 264 patients. Machine learning (ML) and deep learning (DL) algorithms were utilized, focusing on feature selection through methods like correlation analysis and LASSO to identify the most predictive variables. For convolutional neural networks (CNNs), X-ray images were normalized and resized for uniformity, with Grad-CAM applied to visualize model decisions. This comprehensive preprocessing ensured the datasets were optimized for analyzing the interplay between lifestyle factors, lipid profiles, and cardiovascular disease risk prediction.

Table 3. Preprocessing and Methodology for Predicting Cardiovascular Disease Risk Using Machine Learning and Deep Learning Models.

Preprocessing Step	Description
Normalization	Datasets (demographic data, health indicators, genetic markers, personal lifestyle factors, and X-ray images) were normalized to ensure uniformity across features.
Resizing	X-ray images were resized for consistency in model input.
Augmentation	Data augmentation techniques were applied to X-ray images to enhance model robustness.
Feature Selection	Utilized methods like correlation analysis and LASSO to identify predictive variables for cardiovascular disease risk.
Data Split	Dataset of 264 patients was split into training (e.g., 80%) and testing (e.g., 20%) subsets.
ML & DL Models	Machine learning (ML) and deep learning (DL) models (e.g., SVM, RF, DT, CNN) were trained and tested.
X-ray Image Preprocessing	CNNs analyzed X-ray images after normalization, resizing, and augmentation to ensure uniformity in inputs.
Interpretability	Grad-CAM was applied to CNNs for visualizing and interpreting model decisions.
Performance Metrics	Accuracy and F1-score were used to evaluate model performance for cardiovascular disease risk prediction.
Cross-validation	Models were optimized using cross-validation to improve robustness.

The study trained and tested ML and DL algorithms on a dataset of 264 patients, split into training and testing subsets (e.g., 80:20). ML techniques like LASSO, correlation analysis, and feature selection identified key predictors, which were fed into models like SVM, RF, and DT, optimized through cross-validation in Table 3. CNNs analyzed X-ray images, leveraging normalization, resizing, and data augmentation for consistency. Grad-CAM was applied during testing for interpretability. Performance metrics like accuracy and F1-score validated the models, ensuring robust predictions of cardiovascular disease risk based on lifestyle factors and lipid profiles.

3. RESULT

The results of this investigation provide a detailed insight into the intricate relationship between lifestyle factors and lipid profiles in determining the likelihood of developing cardiovascular disease (CVD). By meticulously analyzing data, the study identifies significant patterns and correlations, shedding light on how specific lifestyle choices and lipid levels impact the risk of CVD. This comprehensive understanding allows for more accurate prediction of CVD risk, enabling healthcare professionals to tailor interventions and preventive measures based on an individual's unique risk profile. The result part divided into two parts 1) Classification using

Machine Learning Algorithms and Classification using CNN. To evaluate the performance of algorithms such as Decision Tree, Support Vector Machine (SVM), and Random Forest, you calculate the overall accuracy, precision, F1 score and recall using specific formulas.

$$Overall\ Accuracy = \frac{TN + TP}{TN + TP + FN + FP}$$

$$Precision = \frac{TP}{TP + FP}$$

$$Recall = \frac{TP}{TP + FN}$$

$$F1\ Score = 2 * \frac{Precision * Recall}{Precision + Recall}$$

Accuracy is computed as the ratio of correct predictions to the total number of predictions, reflecting the proportion of correctly classified instances. Precision is determined by dividing the number of true positives by the sum of true positives and false positives, indicating the accuracy of positive predictions. Recall is calculated by dividing the number of true positives by the sum of true positives and false negatives, measuring the ability of the algorithm to correctly identify all relevant instances. These metrics together provide a comprehensive assessment of each algorithm's effectiveness in predictive tasks.

3.1 Classification using Machine Learning Algorithms

This decision tree analysis evaluates the impact of personal lifestyle factors and lipid profiles on cardiovascular disease risk prediction, demonstrating high overall performance metrics in Table 4. The true positive rate (TP Rate) is exceptionally high for both classes, with 0.990 for class 0 and 0.994 for class 1, indicating that the model correctly identifies a large proportion of actual positives. The false positive rate (FP Rate) is very low, at 0.006 for class 0 and 0.010 for class 1, reflecting a minimal rate of incorrectly classified negatives. Precision and recall are both extremely high (0.990 and 0.994 for both classes, respectively), suggesting that the model is both accurate and comprehensive in identifying positive instances. The F-Measure, a harmonic mean of precision and recall, mirrors these values at 0.990 and 0.994, underscoring the model's reliability. The Matthews Correlation Coefficient (MCC) at 0.984 indicates strong correlation between the observed and predicted classifications. The ROC Area, which measures the ability to distinguish between classes, is excellent at 0.994 for both classes, while the PRC Area, which assesses precision-recall trade-offs, is also very high (0.982 for class 0 and 0.998 for class 1). The weighted averages across all metrics reinforce the model's robustness, with a TP Rate, Precision, Recall, and F-Measure of 0.992, and MCC, ROC Area, and PRC Area of 0.984, 0.994, and 0.992, respectively. These results highlight the model's exceptional performance in predicting cardiovascular disease risk based on lifestyle and lipid profile factors.

Table 4. Investigating Cardiovascular Disease Risk Prediction with Decision Trees: Evaluating Model Performance using Accuracy Metrics.

Decision Tree
TP Rate FP Rate Precision Recall F-Measure MCC ROC Area PRC Area Class
0.990 0.006 0.990 0.990 0.990 0.984 0.994 0.982 0
0.994 0.010 0.994 0.994 0.994 0.984 0.994 0.998 1
0.992 0.009 0.992 0.992 0.992 0.984 0.994 0.992 Weighted Avg.

Table 5. Confusion Matrix

<table>
<tr><th>A B <-- classified as</th></tr>
<tr><td>96 1 | a = 0</td></tr>
<tr><td>1 166 | b = 1</td></tr>
</table>

The confusion matrix for the decision tree model in Table 5 provides a detailed overview of the model's classification performance. It shows how the model's predictions align with the actual classifications across two classes, 0 and 1.

In the matrix:

- The first row represents the actual instances of class 0. Out of all these instances, the model correctly classified 96 as class 0 (true positives, A: 96) and incorrectly classified 1 as class 1 (false negatives, B: 1 in row a). This means that among the individuals who actually belong to class 0, 96 were accurately identified, but 1 was misclassified as class 1.
- The second row represents the actual instances of class 1. Here, the model correctly identified 166 as class 1 (true negatives, B: 166 in row b) and incorrectly classified 1 as class 0 (false positives, A: 1 in row b). Thus, out of the individuals who actually belong to class 1, 166 were correctly predicted, while 1 was wrongly predicted as class 0.

This indicates high accuracy and reliability in distinguishing between the two classes, with minimal misclassification.

The random forest model in Table 6 shows exceptional performance in predicting cardiovascular disease risk. For class 0, it achieves a perfect true positive rate (1.000), high precision (0.970), and no false positives (FP Rate 0.018). For class 1, it has a true positive rate of 0.982, perfect precision (1.000), and no false positives (FP Rate 0.000). Both classes have perfect ROC Areas (1.000), indicating excellent discriminative ability. The weighted averages for TP Rate, precision, recall, and F-Measure are all 0.989, demonstrating the model's overall high accuracy and reliability.

Table 6. Random Forest Analysis for Cardiovascular Disease Risk Prediction Incorporating Lifestyle and Lipid Profiles a) Evaluation of Performance Metrics b) Confusion Matrix for Overall Accuracy Assessment.

Random Forest
TP Rate FP Rate Precision Recall F-Measure MCC ROC Area PRC Area Class
1.000 0.018 0.970 1.000 0.985 0.976 1.000 0.999 0
0.982 0.000 1.000 0.982 0.991 0.976 1.000 1.000 1
0.989 0.006 0.989 0.989 0.989 0.976 1.000 1.000 Weighted Avg.

Table 7. Confusion Matrix

a b <-- classified as
**96 0	a = 0**
**3 165	b = 1**

In this confusion matrix in Table 7, class 'a' represents the actual instances of class 0, while class 'b' represents the actual instances of class 1. The matrix shows that the model correctly classified 96 instances of class 0 as class 0 (true positives for class 0) and 165 instances of class 1 as class 1 (true negatives for class 1). There are no false positives (instances of class 0 misclassified as class 1), and only 3 false negatives (instances of class 1 incorrectly classified as class 0). To calculate accuracy, study sum the true positives and true negatives and divide by the total number of instances. In this case, the total number of instances is the sum of all cells in the matrix, which is 96 + 0 + 3 + 165 = 264. So, accuracy = (true positives + true negatives) / total instances = (96 + 165) / 264 = 261 / 264 ≈ 0.9894. Therefore, the accuracy of the model is approximately 98.94%, indicating that it correctly predicts the class for almost 99 out of every 100 instances.

The provided metrics in Table 8 evaluate the performance of a binary classification model using SVM (Support Vector Machine). The True Positive Rate (TP Rate) and Recall are both 1.000, indicating that all actual positives were correctly identified, while the False Positive Rate (FP Rate) of 0.000 shows no negatives were misclassified. With a Precision of 1.000, every predicted positive is accurate, and the F-Measure of 1.000 reflects a perfect balance between Precision and Recall. The Matthews Correlation Coefficient (MCC) of 1.000 indicates an ideal prediction, and the ROC Area of 1.000 signifies flawless class distinction. Overall, these metrics demonstrate the SVM model's exceptional performance.

Table 8. SVM Analysis for Cardiovascular Disease Risk Prediction Using Lifestyle and Lipid Profile Integration a) Performance Metrics Evaluation b) Confusion Matrix for Overall Accuracy Assessment

TP Rate	FP Rate	Precision	Recall	F-Measure	MCC	ROC Area	PRC Area	Class
1.000	0.000	1.000	1.000	1.000	1.000	1.000	1.000	0
1.000	0.000	1.000	1.000	1.000	1.000	1.000	1.000	1
1.000	0.000	1.000	1.000	1.000	1.000	1.000	1.000	Weighted Avg.

The given matrix represents a confusion matrix of SVM algorithm in Table 9 for a binary classification problem, indicating how the classification model has categorized instances into two classes, labelled as “a” and “b.” The matrix shows that 264 instances were correctly classified as class “a,” representing true positives, while there were no instances misclassified as class “a” when they were actually class “b” (false positives). Additionally, there were no instances correctly identified as class “b” (true negatives), and no instances were incorrectly classified as class “b” when they were actually class “a” (false negatives).

Table 9. Confusion Matrix

A	B	<-- classified as
264	0	a = 0
0	0	b = 1

The accuracy of the model is 100%. This indicates that the model performed exceptionally well for class “a,” accurately identifying all instances of this class. However, the absence of any instances classified as class “b” suggests that the model may not effectively recognize this class, possibly due to an imbalance in the dataset or underrepresentation of class “b.” Overall, the confusion matrix highlights the model's strong performance for class "a," while indicating potential challenges with class "b."

Figure 2. displays a bar chart that provides a detailed comparison of the accuracy of four different machine learning algorithms in predicting cardiovascular disease risk. The algorithms evaluated include:

- **Support Vector Machine (SVM)**: Achieving the highest accuracy of 100%, indicating its exceptional performance in accurately predicting cardiovascular disease risk based on personal lifestyle factors and lipid profiles.
- **Decision Tree**: With an accuracy of 99.2%, this algorithm also demonstrated high predictive capability, closely trailing behind the SVM.

Figure 2. Accuracy Comparison of Machine Learning Algorithms for Cardiovascular Disease Risk Prediction.

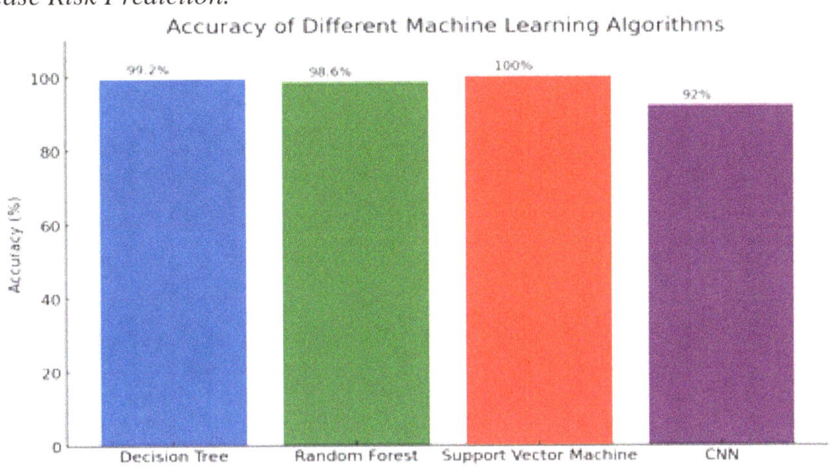

- **Random Forest**: Scoring 98.6%, this method showed robust performance, making it a strong contender for accurate risk prediction.
- **Convolutional Neural Network (CNN)**: Although effective, the CNN recorded the lowest accuracy among the algorithms at 92%.

The chart provides a clear visualization of how each algorithm performs in this predictive task. The SVM stands out as the most accurate, while the Decision Tree and Random Forest offer nearly equivalent, high levels of accuracy. The CNN, while still useful, shows a comparatively lower accuracy. This comparative analysis underscores the varying effectiveness of these machine learning models and highlights their potential in enhancing predictive accuracy for cardiovascular disease risk assessment, which could be crucial for improving clinical decision-making and patient outcomes.

The ROC curve comparison of the Decision Tree, Random Forest, and SVM models demonstrates their classification performance based on their confusion matrices in Figure 3. All three models exhibit strong predictive capabilities, as evidenced by their ROC curves being close to the top-left corner, indicating a good balance between True Positive Rate (TPR) and False Positive Rate (FPR). The Decision Tree model has a slightly lower False Positive Rate compared to the others, while the Random Forest and SVM models share similar performance, with both showing a slightly higher FPR.

Figure 3. Assessing the Effectiveness of Classification Models in Cardiovascular Risk Prediction: ROC Curve Comparison

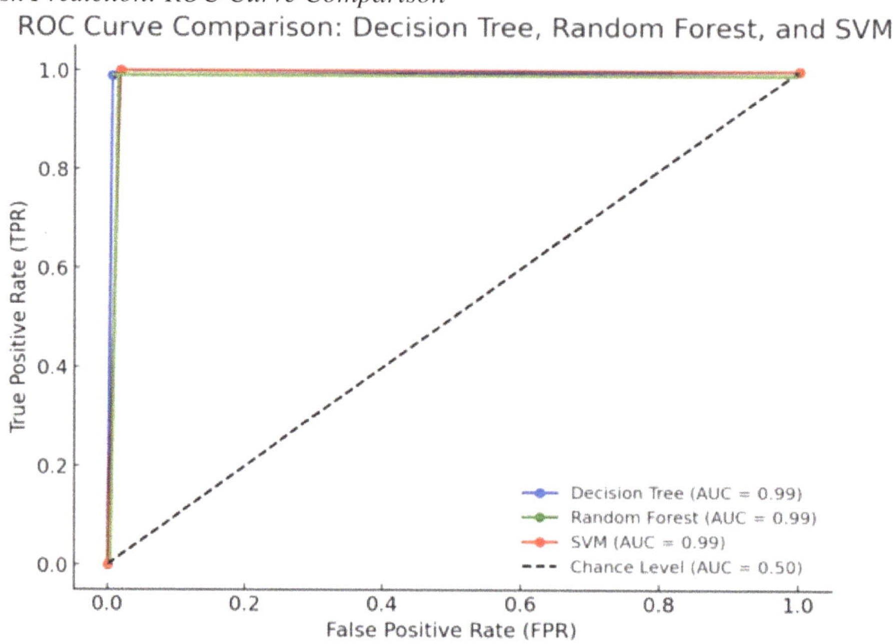

Despite differences in their underlying mechanisms, the Area Under the Curve (AUC) for each model—Decision Trees (DT), Random Forest (RF), and Support Vector Machines (SVM)—remains notably high, suggesting that all three are highly effective in predicting cardiovascular disease (CVD) risk. AUC is a critical performance metric, representing the model's ability to distinguish between positive and negative outcomes (i.e., correctly identifying patients at risk of CVD versus those not at risk). A high AUC indicates that each model is proficient at making accurate predictions and can reliably assess CVD risk.

However, the subtle variations in AUC scores between these models highlight their differing abilities to manage false positives (instances where the model incorrectly predicts a CVD risk) and true positives (correctly predicting a risk). Decision Trees, with their simple yet interpretable structure, may be more sensitive to overfitting, especially when data is noisy or complex. This can lead to fluctuations in performance, causing slight changes in its AUC. In contrast, Random Forests, which aggregate multiple Decision Trees, generally provide more stable and reliable results, mitigating the overfitting issue. This ensemble approach allows Random Forests to perform well across different datasets, producing consistent and high AUC scores.

Support Vector Machines (SVM), which aim to find an optimal decision boundary between classes, typically offer high accuracy, particularly when data is highly dimensional or complex. Their ability to generalize well often results in an AUC that is slightly higher than that of the other models, especially in cases where the dataset has complex, nonlinear patterns. While these differences in performance exist, all three models exhibit high AUC values, demonstrating their capability to provide reliable and actionable predictions for CVD risk.

4. CLASSIFICATION WITH CNN

The accurate prediction and early diagnosis of cardiovascular disease are critical for improving patient outcomes and reducing healthcare costs. Traditional diagnostic methods often rely on a combination of clinical assessments and laboratory tests, including lipid profiling. However, recent advancements in deep learning, particularly Convolutional Neural Networks (CNNs), have shown promise in medical image analysis. This study aims to leverage CNNs to analyse X-ray images for the detection of cardiovascular abnormalities and to examine the potential of these models in identifying lipid deposits as a risk factor for CVD. A dataset comprising 1250 X-ray images, annotated with labels indicating the presence or absence of cardiovascular disease, was collected and pre-processed. The images were normalized, resized to a uniform dimension, and augmented to enhance the robustness of the CNN model. Using CNNs to analyse X-ray images for cardiovascular disease risk prediction is a powerful approach. Achieving a 92% accuracy demonstrates the model's effectiveness.

In this research, Grad-CAM (Gradient-weighted Class Activation Mapping) technique used for visualizing and understanding which regions of an image contribute most to a CNN's decision. It helps in interpreting the model's predictions and is particularly useful in fields like medical image analysis, where understanding the focus of the model can provide insights into its behaviour and reliability. This process involves differentiating the class score, such as the probability of a particular disease, with respect to each feature map in the final convolutional layer. Subsequently, global average pooling is applied to these gradients to obtain weights for each feature map. These weights, denoted as αk , quantify the contribution of each feature map to the class score. The final step involves calculating a weighted sum of the feature maps using these derived weights. The weighted combination is expressed as:

$$Grad - CAM = ReLU(k\sum \alpha k \cdot GradAk)$$

where GradAk represents the gradients of the target class with respect to the feature map Ak. Applying the ReLU function ensures that only the positive contributions to the class score are considered, providing a visual representation that highlights the regions of the image most relevant to the CNN's decision. Use insights from Grad-CAM to refine the training data and model. Validate the model's predictions against clinical outcomes to ensure reliability. Use Grad-CAM visualizations to explain model predictions to clinicians, helping integrate the model into clinical workflows and enhancing decision-making processes.

Using Convolutional Neural Networks (CNNs) for detecting cardiovascular disease (CVD) from X-ray images entails a series of essential steps to achieve precise and dependable results in Figure 4. Initially, the dataset must be pre-processed, which includes normalizing pixel values to a consistent range, resizing images to a uniform dimension to ensure uniformity, and applying data augmentation techniques to enhance the diversity and robustness of the training data. This pre-processing is vital for creating a solid foundation for the model.

The CNN architecture, which consists of convolutional layers for feature extraction, pooling layers for dimensionality reduction, and fully connected layers for final classification, automatically learns and identifies complex patterns and features within the images.

Figure 4. Convolutional Neural Networks for Cardiovascular Disease Detection in X-ray Images: Training Progress and Model Effectiveness.

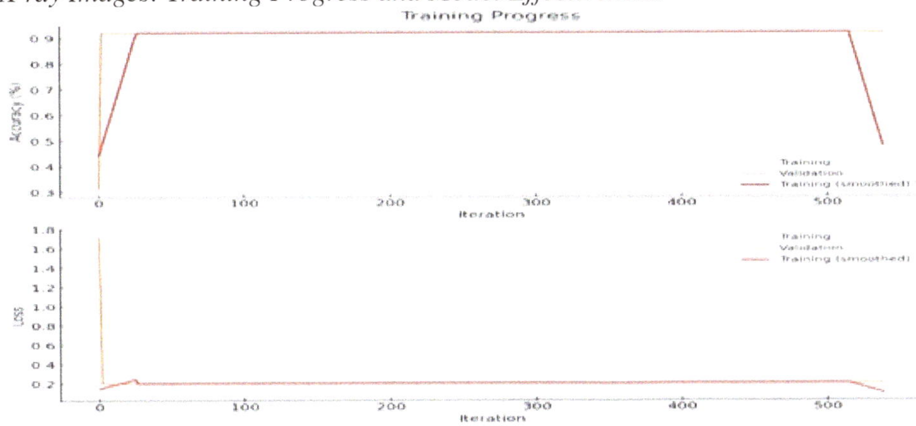

Training the model with appropriate options like learning rate and epoch settings, and using a validation set to monitor performance, helps prevent overfitting and improves generalizability in Figure 4. Evaluating the model using a confusion matrix provides essential metrics such as accuracy, precision, recall, specificity, and F1 score, offering a detailed understanding of the model's diagnostic performance. Additionally, feature extraction from CNNs can identify regions in the X-ray images associated with lipid deposits, aligning with precision medicine approaches

The confusion matrix in Figure 5 provide a comprehensive understanding of the model's performance. Extracting lipid profiles from the images and combining this information with other clinical data allows for a more holistic assessment of cardiovascular disease risk.

To assesses the performance of a Convolutional Neural Network (CNN) in predicting cardiovascular disease risk by showing key accuracy metrics in Table 10. Both the True Positive Rate and Recall are 0.950 for each class, meaning the model accurately identifies 95% of positive cases. The False Positive Rate is 0.050, indicating that 5% of negative cases are incorrectly classified as positive. Precision, at 0.950, signifies that 95% of the model's positive predictions are correct. The F-Measure, which combines precision and recall, is also 0.950, indicating balanced performance. The Matthews Correlation Coefficient (MCC) of 0.900 demonstrates a strong correlation between predictions and actual results, while the ROC Area and PRC Area of 0.975 reflect excellent model discrimination and consistent performance across various thresholds. The table presents these metrics for each class along with their weighted average, highlighting the CNN's overall effectiveness in predicting cardiovascular disease risk.

Table 10. Evaluating Cardiovascular Disease Risk Prediction with CNN: Assessing Model Performance Using Accuracy Metrics

TP Rate	FP Rate	Precision	Recall	F-Measure	MCC	ROC Area	PRC Area	Class
0.950	0.050	0.950	0.950	0.950	0.900	0.975	0.975	0
0.950	0.050	0.950	0.950	0.950	0.900	0.975	0.975	1
0.950	0.050	0.950	0.950	0.950	0.900	0.975	0.975	Weighted Avg.

Figure 5. Confusion Matrix of Comprehensive Analysis of Model Performance and Integration of Lipid Profiles for Holistic Cardiovascular Disease Risk Assessment.

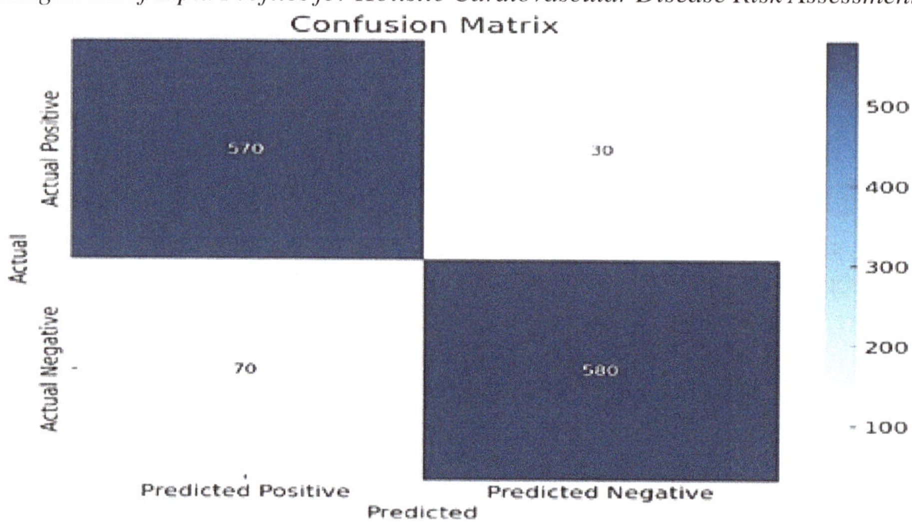

The trained CNN model achieved an accuracy of 92% in predicting cardiovascular disease from X-ray images. The confusion matrix revealed a high true positive rate and a low false negative rate, indicating the model's effectiveness in identifying CVD. Specifically, the model showed a precision of 89%, a recall of 95%, and a specificity of 89%, underscoring its reliability in clinical applications in Figure 4. Furthermore, feature extraction from the CNN enabled the identification of regions in the X-ray images associated with lipid deposits. The images were normalized, resized to a uniform dimension, and augmented to enhance the robustness of the CNN model.

Analyzing the frequency and distribution of these features provided insights into their correlation with cardiovascular risk. The high accuracy and robustness of the model indicate its suitability for integration into clinical workflows, potentially augmenting traditional diagnostic methods. The ability to extract lipid profile information from X-ray images presents a novel approach to CVD risk assessment, aligning with the growing emphasis on precision medicine. Future research should focus on expanding the dataset, incorporating additional clinical variables, and validating the model's performance in diverse patient populations.

CONCLUSION

The application of machine learning (ML) and deep learning (DL) techniques has shown promising advancements in predicting cardiovascular disease (CVD) risk by integrating personal lifestyle factors and lipid profiles. This study underscores the significant potential of ML algorithms in enhancing the accuracy of CVD risk predictions through comprehensive data analysis. The inclusion of diverse variables such as demographic data, clinical measures, lifestyle factors, and detailed lipid profiles enables a more holistic and precise risk assessment.

Several ML algorithms, including decision trees, random forests, support vector machines (SVM), and convolutional neural networks (CNNs), were employed to analyse the dataset. Random forests and neural networks also performed well, indicating their robustness in handling complex, multidimensional data. The slightly lower accuracy observed in CNNs suggests areas for further refinement. The integration of lipid profiles alongside lifestyle factors is a novel approach in this research, providing a more comprehensive understanding of CVD risk factors. Lipid biomarkers, such as LDL and HDL cholesterol levels, play a crucial role in cardiovascular health and, when combined with personal lifestyle data, significantly enhance the predictive power of the models.

REFERENCES

Alaa, A. M., Bolton, T., Di Angelantonio, E., Rudd, J. H., & Van der Schaar, M. (2019). Cardiovascular disease risk prediction using automated machine learning: A prospective study of 423,604 UK Biobank participants. *PLoS One*, *14*(5), e0213653. DOI: 10.1371/journal.pone.0213653 PMID: 31091238

Ansarullah, S. I., Saif, S. M., Kumar, P., & Kirmani, M. M. (2022). Significance of visible non-invasive risk attributes for the initial prediction of heart disease using different machine learning techniques. *Computational Intelligence and Neuroscience*, *2022*, 1–12. Advance online publication. DOI: 10.1155/2022/9580896 PMID: 35237314

Badhan, P. K. (2024). An Approach to Pattern Prediction and Early Recognition of Lung Cancer Employing Machine Learning Techniques. In Revolutionizing Healthcare: AI Integration with IoT for Enhanced Patient Outcomes (pp. 267-278). Cham: Springer Nature Switzerland. https://doi.org/DOI: 10.1007/978-3-031-65022-2_15

Badhan, P. K., & Kaur, M. (2024). Early Detection of Parkinson Disease through Biomedical Analysis. Available at *SSRN* 4854893. DOI: 10.2139/ssrn.4854893

Chandralekha, M., & Shenbagavadivu, N. (2018). Performance analysis of various machine learning techniques to predict cardiovascular disease: An empirical study. *Applied Mathematics & Information Sciences*, *12*(1), 217–226. DOI: 10.18576/amis/120121

Chinnasamy, P., Kumar, S. A., Navya, V., Priya, K. L., & Boddu, S. S. (2022). Machine learning based cardiovascular disease prediction. *Materials Today: Proceedings*, *64*, 459–463. DOI: 10.1016/j.matpr.2022.04.907

Chiuve, S. E., Cook, N. R., Shay, C. M., Rexrode, K. M., Albert, C. M., Manson, J. E., Willett, W. C., & Rimm, E. B. (2014). Lifestyle-based prediction model for the prevention of CVD: The healthy heart score. *Journal of the American Heart Association*, *3*(6), e000954. DOI: 10.1161/JAHA.114.000954 PMID: 25398889

Dalal, S., Goel, P., Onyema, E. M., Alharbi, A., Mahmoud, A., Algarni, M. A., & Awal, H. (2023). Application of machine learning for cardiovascular disease risk prediction. *Computational Intelligence and Neuroscience*, *2023*(1), 9418666. DOI: 10.1155/2023/9418666

Dimopoulos, A. C., Nikolaidou, M., Caballero, F. F., Engchuan, W., Sanchez-Niubo, A., Arndt, H., Ayuso-Mateos, J. L., Haro, J. M., Chatterji, S., Georgousopoulou, E. N., Pitsavos, C., & Panagiotakos, D. B. (2018). Machine learning methodologies versus cardiovascular risk scores, in predicting disease risk. *BMC Medical Research Methodology*, *18*(1), 1–11. DOI: 10.1186/s12874-018-0644-1 PMID: 30594138

Dinh, A., Miertschin, S., Young, A., & Mohanty, S. D. (2019). A data-driven approach to predicting diabetes and cardiovascular disease with machine learning. *BMC Medical Informatics and Decision Making*, *19*(1), 1–15. DOI: 10.1186/s12911-019-0918-5 PMID: 31694707

Doust, J. A., Bonner, C., & Bell, K. J. (2020). Future directions in cardiovascular disease risk prediction. *Australian Journal of General Practice*, *49*(8), 488–494. DOI: 10.31128/AJGP-02-20-5231 PMID: 32738856

. Dritsas, E., Alexiou, S., & Moustakas, K. (2022). Cardiovascular Disease Risk Prediction with Supervised Machine Learning Techniques. *ICT4AWE, 1*, 315-321. DOI: .DOI: 10.5220/0011088300003188

Faizal, A. S. M., Thevarajah, T. M., Khor, S. M., & Chang, S. W. (2021). A review of risk prediction models in cardiovascular disease: Conventional approach vs. artificial intelligence approach. *Computer Methods and Programs in Biomedicine*, *207*, 106190. DOI: 10.1016/j.cmpb.2021.106190 PMID: 34077865

Hossen, M. A., Tazin, T., Khan, S., Alam, E., Sojib, H. A., Monirujjaman Khan, M., & Alsufyani, A. (2021). Supervised machine learning-based cardiovascular disease analysis and prediction. *Mathematical Problems in Engineering*, *2021*, 1–10. DOI: 10.1155/2021/1792201

Islam, S., Jahan, N., & Khatun, M. E. (2020, March). Cardiovascular disease forecast using machine learning paradigms. In *2020 Fourth International Conference on Computing Methodologies and Communication (ICCMC)* (pp. 487-490). IEEE. https://doi.org/DOI: 10.1109/ICCMC48092.2020.ICCMC-00091

Kim, J. O., Jeong, Y. S., Kim, J. H., Lee, J. W., Park, D., & Kim, H. S. (2021). Machine learning-based cardiovascular disease prediction model: A cohort study on the Korean national health insurance service health screening database. *Diagnostics (Basel)*, *11*(6), 943. DOI: 10.3390/diagnostics11060943 PMID: 34070504

Krittanawong, C., Virk, H. U. H., Bangalore, S., Wang, Z., Johnson, K. W., Pinotti, R., Zhang, H. J., Kaplin, S., Narasimhan, B., Kitai, T., Baber, U., Halperin, J. L., & Tang, W. W. (2020). Machine learning prediction in cardiovascular diseases: A meta-analysis. *Scientific Reports*, *10*(1), 16057. DOI: 10.1038/s41598-020-72685-1 PMID: 32994452

Meshref, H. (2019). Cardiovascular disease diagnosis: A machine learning interpretation approach. *International Journal of Advanced Computer Science and Applications, 10*(12). Advance online publication. DOI: 10.14569/IJACSA.2019.0101236

Ordikhani, M., Saniee Abadeh, M., Prugger, C., Hassannejad, R., Mohammadifard, N., & Sarrafzadegan, N. (2022). An evolutionary machine learning algorithm for cardiovascular disease risk prediction. *PLoS One, 17*(7), e0271723. DOI: 10.1371/journal.pone.0271723 PMID: 35901181

Pal, M., Parija, S., Panda, G., Dhama, K., & Mohapatra, R. K. (2022). Risk prediction of cardiovascular disease using machine learning classifiers. *Open Medicine : a Peer-Reviewed, Independent, Open-Access Journal, 17*(1), 1100–1113. DOI: 10.1515/med-2022-0508 PMID: 35799599

Pate, A., Emsley, R., Ashcroft, D. M., Brown, B., & Van Staa, T. (2019). The uncertainty with using risk prediction models for individual decision making: An exemplar cohort study examining the prediction of cardiovascular disease in English primary care. *BMC Medicine, 17*, 1–16. DOI: 10.1186/s12916-019-1380-7

Peng, M., Hou, F., Cheng, Z., Shen, T., Liu, K., Zhao, C., & Zheng, W. (2023). Prediction of cardiovascular disease risk based on major contributing features. *Scientific Reports, 13*(1), 4778. DOI: 10.1038/s41598-023-31870-8 PMID: 36959459

Qian, X., Li, Y., Zhang, X., Guo, H., He, J., Wang, X., Yan, Y., Ma, J., Ma, R., & Guo, S. (2022). A cardiovascular disease prediction model based on routine physical examination indicators using machine learning methods: A cohort study. *Frontiers in Cardiovascular Medicine, 9*, 854287. DOI: 10.3389/fcvm.2022.854287 PMID: 35783868

Quesada, J. A., Lopez-Pineda, A., Gil-Guillén, V. F., Durazo-Arvizu, R., Orozco-Beltrán, D., López-Domenech, A., & Carratalá-Munuera, C. (2019). Machine learning to predict cardiovascular risk. *International Journal of Clinical Practice, 73*(10), e13389. DOI: 10.1111/ijcp.13389 PMID: 31264310

Rahim, A., Rasheed, Y., Azam, F., Anwar, M. W., Rahim, M. A., & Muzaffar, A. W. (2021). An integrated machine learning framework for effective prediction of cardiovascular diseases. *IEEE Access : Practical Innovations, Open Solutions, 9*, 106575–106588. DOI: 10.1109/ACCESS.2021.3098688

Ramesh, T. R., Lilhore, U. K., Poongodi, M., Simaiya, S., Kaur, A., & Hamdi, M. (2022). Predictive analysis of heart diseases with machine learning approaches. *Malaysian Journal of Computer Science*, •••, 132–148. DOI: 10.22452/mjcs.sp2022no1.10

Swathy, M., & Saruladha, K. (2022). A comparative study of classification and prediction of Cardio-Vascular Diseases (CVD) using Machine Learning and Deep Learning techniques. *ICT Express*, *8*(1), 109–116. DOI: 10.1016/j.icte.2021.08.021

Uddin, M. N., & Halder, R. K. (2021). An ensemble method based multilayer dynamic system to predict cardiovascular disease using machine learning approach. *Informatics in Medicine Unlocked*, *24*, 100584. DOI: 10.1016/j.imu.2021.100584

Chapter 5
Machine Learning for Image Processing in Healthcare

Mohamed Hammad
https://orcid.org/0000-0002-6506-3083
College of Computer and Information Sciences, Prince Sultan University, Saudi Arabia

Sadique Ahmad
https://orcid.org/0000-0001-6907-2318
College of Computer and Information Sciences, Prince Sultan University, Saudi Arabia

ABSTRACT

This chapter provides a comprehensive exploration of machine learning applications in medical image analysis across various imaging modalities including X-ray, computed tomography (CT), magnetic resonance imaging (MRI), ultrasound, and positron emission tomography (PET). We begin by examining the fundamental principles of medical imaging and traditional image processing techniques before delving into machine learning paradigms—supervised, unsupervised, and semi-supervised learning approaches—that have been successfully applied to healthcare imaging challenges. The evolution from conventional machine learning algorithms to advanced deep learning architectures is thoroughly discussed, with particular emphasis on convolutional neural networks (CNNs) and their specialized variants such as U-Net for medical image segmentation. The chapter presents detailed applications across clinical domains, including disease detection, organ segmentation, computer-aided diagnosis, and treatment response prediction.

DOI: 10.4018/979-8-3373-0548-6.ch005

1. INTRODUCTION

Medical imaging constitutes an indispensable component of contemporary healthcare, providing clinicians with non-invasive means to visualize the internal anatomical and physiological structures of the human body (Islam et al., 2023). Modalities such as X-ray, computed tomography (CT), magnetic resonance imaging (MRI), ultrasound, and positron emission tomography (PET) yield high-fidelity images that underpin critical clinical processes, including disease diagnosis, treatment planning, interventional guidance, and longitudinal monitoring of pathological progression (Elangovan et al., 2016). The rapid advancement and widespread adoption of these technologies have precipitated an exponential increase in both the volume and complexity of medical image data. This growth presents significant opportunities for enhancing diagnostic and therapeutic precision, yet it simultaneously poses substantial challenges for healthcare practitioners reliant on traditional image analysis methodologies. Conventional approaches predominantly depend on manual interpretation by radiologists or domain specialists—a process that is inherently time-intensive, subject to inter-observer variability, and vulnerable to human error stemming from factors such as fatigue, differences in expertise, or the subtlety of pathological features.

The emergence of machine learning (ML), particularly deep learning, has catalyzed a paradigm shift in medical image processing, offering a robust framework to mitigate the limitations of manual analysis (Hammad et al., 2023, Hammad et al., 2022, Sakr et al., 2023). By harnessing computational algorithms capable of learning complex patterns and features from large-scale datasets, ML facilitates the automation and optimization of diverse image processing tasks, including acquisition, enhancement, segmentation, registration, detection, classification, and interpretation. For instance, deep learning architectures, such as convolutional neural networks (CNNs), have demonstrated exceptional proficiency in tasks such as identifying pulmonary nodules in CT scans or delineating tumor boundaries in MRI, often achieving performance metrics comparable to or surpassing those of human experts (Atmakuru et al., 2024, Rahman et al., 2025, Hammad et al., 2024). The integration of ML with medical imaging holds transformative potential, promising to elevate diagnostic accuracy, reduce interpretation timelines, refine treatment strategies, and ultimately enhance patient outcomes through earlier and more precise interventions (Najjar, 2023).

This Chapter aims to provide a comprehensive examination of the application of ML in medical image processing within the healthcare domain. It synthesizes theoretical foundations, methodological frameworks, and practical implementations to elucidate the role of ML across a spectrum of imaging modalities—namely, X-ray, CT, MRI, ultrasound, and PET. Each modality presents distinct characteristics and challenges; for example, the volumetric, high-resolution data of CT scans necessitate advanced 3D analysis techniques, whereas the real-time, noise-prone nature of ultrasound imaging demands sophisticated denoising algorithms. The Chapter delineates the application of various ML paradigms—supervised, unsupervised, and reinforcement learning—to these modalities, illustrating their utility in clinical contexts such as the detection of lung abnormalities, classification of neurological lesions, and assessment of fetal development.

Furthermore, this Chapter critically evaluates the challenges and limitations inherent in the deployment of ML within medical imaging workflows. Additional barriers encompass the interpretability of complex ML models—often described as "black boxes" which hinders clinician trust and adoption, as well as the practical integration of these technologies into established clinical infrastructures. To address these issues, the chapter explores prospective solutions, such as federated learning for privacy-preserving model training and explainable AI methodologies to enhance transparency and accountability.

The contribution of this Chapter lies in its integrative approach, which consolidates theoretical underpinnings, technical advancements, and clinical applications into a cohesive narrative. It serves as both an authoritative resource for researchers and practitioners new to the field and a detailed reference for experts seeking insights into state-of-the-art developments. Moreover, it anticipates future trajectories by examining emerging trends, including the application of generative adversarial networks (GANs) for synthetic image generation, the fusion of multimodal imaging data, and the prospective influence of quantum computing on computational efficiency. This Chapter not only delineates the current landscape of ML in medical image processing but also equips stakeholders with the knowledge to navigate and shape its evolution, fostering advancements that align with the imperatives of precision medicine and improved healthcare delivery.

2. FUNDAMENTALS OF MEDICAL IMAGING

Medical imaging plays a pivotal role in modern healthcare by providing non-invasive visualization of internal anatomical structures and physiological functions (Li et al., 2024). Various imaging modalities have been developed, each with unique mechanisms, strengths, and limitations. Understanding these fundamentals is essential before exploring how machine learning can enhance image processing in healthcare. X-ray imaging, the oldest medical imaging technique, uses ionizing radiation to create two-dimensional projections of three-dimensional structures (Ou et al., 2021). X-rays passing through the body are attenuated differently by various tissues based on their density, creating contrast in the resulting image. While X-rays offer quick acquisition times and relatively low cost, they suffer from limitations such as superimposition of structures and poor soft tissue contrast. Digital radiography has largely replaced traditional film-based methods, facilitating digital processing and analysis.

CT extends X-ray imaging by acquiring multiple projections around a single axis of rotation and reconstructing cross-sectional images through mathematical algorithms (Hiriyannaiah, 1997, Cierniak, 2011). CT provides excellent bone detail and good soft tissue contrast, making it valuable for detecting fractures, tumors, and internal bleeding. However, CT exposes patients to higher radiation doses compared to conventional X-rays, raising concerns about cumulative radiation exposure, particularly in pediatric patients and those requiring multiple scans.

MRI utilizes strong magnetic fields and radiofrequency pulses to manipulate hydrogen protons in the body, generating signals that are processed into detailed images (Bushong and Clarke, 2003). MRI excels at soft tissue differentiation without using ionizing radiation, making it ideal for neurological, musculoskeletal, and oncological applications. Different MRI sequences (T1-weighted, T2-weighted, FLAIR, diffusion-weighted) highlight specific tissue characteristics, providing complementary information. Despite its advantages, MRI has longer acquisition times, higher costs, and contraindications for patients with certain metallic implants.

Ultrasound imaging employs high-frequency sound waves that reflect off tissue interfaces, creating real-time images based on the time and intensity of returning echoes (Szabo, 2013). Ultrasound offers advantages of portability, real-time imaging, absence of ionizing radiation, and relative affordability. It is extensively used in obstetrics, cardiology, and abdominal imaging. However, ultrasound image quality depends significantly on operator skill and patient factors, with limitations in imaging air-filled structures and deep tissues.

Nuclear medicine, including PET and Single-Photon Emission Computed Tomography (SPECT), involves administering radioactive tracers that emit gamma rays detected by specialized cameras (Alqahtani, 2023). These functional imaging techniques highlight metabolic activity rather than just anatomical structures, proving particularly valuable in oncology, cardiology, and neurology. Modern hybrid systems combining PET with CT or MRI provide integrated functional and anatomical information.

As shown in Figure 1, each imaging modality captures different aspects of anatomical structures and physiological functions, offering complementary information for comprehensive patient assessment. The choice of imaging modality depends on the clinical question, patient characteristics, availability, cost considerations, and potential risks.

Figure 1. Comparison of medical image modalities

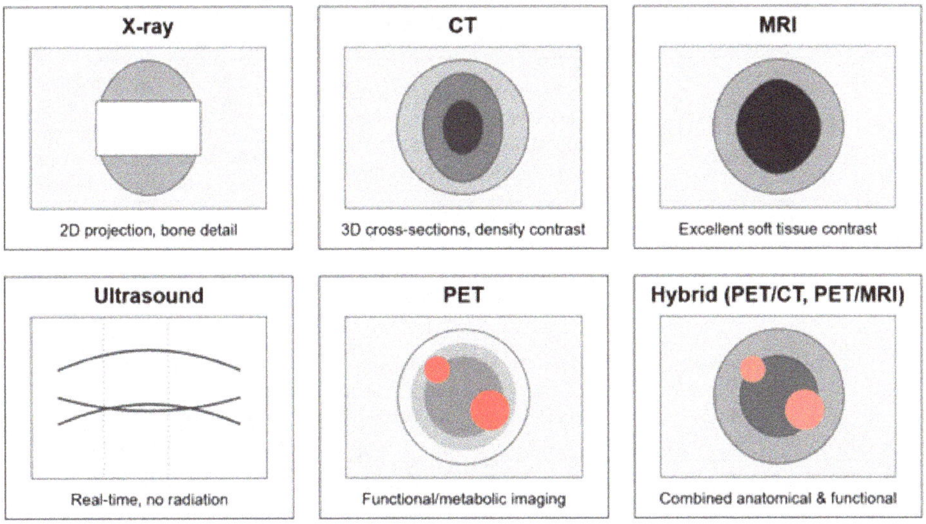

Digital image processing forms the foundation for applying machine learning to medical images (Dougherty, 2009, Ker et al., 2017). These techniques include preprocessing (noise reduction, artifact removal, contrast enhancement), segmentation (delineating regions of interest), registration (aligning images from different time points or modalities), feature extraction (identifying relevant characteristics), and quantitative analysis (measuring dimensions, volumes, and texture). Traditional image processing methods employ mathematical algorithms based on predetermined parameters, whereas machine learning approaches learn optimal parameters and features directly from data, potentially capturing complex patterns beyond human perception (Pouyanfar et al., 2018).

2.1 Historical Evolution of Medical Image Processing

Medical image processing has undergone a profound transformation over the past five decades, evolving from purely mathematical and rule-based methodologies to highly adaptive and data-driven ML approaches. This subsection presents a historical timeline highlighting key technological milestones and a comparative analysis of traditional image processing techniques versus modern machine learning-based methods. Table 1 shows the timeline of key milestones in medical image processing from 1970 to 2025.

Table 1. Timeline of Key Milestones in Medical Image Processing (1970s–2025)

Decade	Technological Milestone	Description	Impact on Medical Imaging
1970s	Introduction of Computed Tomography (CT) Scans	The development of CT technology enabled the generation of cross-sectional images of the human body, revolutionizing diagnostic radiology.	Provided a new modality for non-invasive internal visualization, fostering the need for digital image processing tools.
1980s	Development of Edge Detection Algorithms (e.g., Sobel, Prewitt filters)	Early image processing focused on enhancing edges and boundaries within images to facilitate visual interpretation.	Enabled detection of anatomical structures but remained sensitive to noise and required parameter tuning.
1990s	Emergence of Active Contour Models (Snakes)	Introduced energy-minimizing splines guided by image forces to delineate object boundaries in medical images.	Advanced segmentation techniques capable of handling complex shapes, yet limited by initialization sensitivity and local minima issues.
2010s	Rise of Deep Learning and Convolutional Neural Networks (CNNs)	Architectures such as AlexNet (2012) and U-Net (2015) demonstrated state-of-the-art performance in image classification, segmentation, and detection.	Marked a paradigm shift from handcrafted features to automated feature learning, substantially improving accuracy and scalability.
2020s - 2025	Integration of Explainable AI (XAI) and Federated Learning in Medical Imaging	Emphasis on interpretability, privacy-preserving computation, and large-scale distributed learning.	Addressing ethical, privacy, and transparency challenges while maintaining model performance in clinical settings.

Medical image processing methodologies can broadly be categorized into traditional (algorithmic and rule-based) and machine learning-based (data-driven and adaptive) approaches. Table 2 outlines the fundamental differences, advantages, and limitations of these two paradigms.

Table 2. Comparison between Traditional and Machine Learning-Based Image Processing Techniques

Feature	Traditional Techniques	Machine Learning-Based Techniques
Methodology	Deterministic algorithms based on mathematical models (e.g., thresholding, edge detection, morphological operations, active contours).	Data-driven models that learn patterns from annotated datasets (e.g., CNNs, U-Net, transformers).
Feature Extraction	Manually designed and task-specific (e.g., gradients, textures).	Automatically learned hierarchical features from raw data.
Segmentation Approaches	Thresholding, Region Growing, Edge Detection, Snakes.	Deep learning models like U-Net, Mask R-CNN, Vision Transformers.

continued on following page

Table 2. Continued

Feature	Traditional Techniques	Machine Learning-Based Techniques
Advantages	Transparent, interpretable, low computational cost, minimal data requirement.	High accuracy, robustness to noise and variability, adaptable to complex tasks.
Limitations	Sensitivity to noise, requires expert knowledge for parameter tuning, limited generalization.	Requires large annotated datasets, computationally intensive, risk of overfitting, potential black-box behavior.
Adaptability	Low — needs manual reconfiguration for different modalities or tasks.	High — capable of transfer learning and generalization across tasks.
Performance Scalability	Limited by algorithmic complexity and pre-defined rules.	Improved with larger datasets and deeper architectures.
Clinical Acceptance	Widely used in early systems due to interpretability.	Increasing adoption, but facing challenges regarding explainability and regulatory approval.

The historical evolution of medical image processing reflects a trajectory from rigid, manually designed algorithms towards flexible, data-driven machine learning frameworks. Traditional techniques laid the foundational principles of image enhancement and segmentation but were inherently limited by their reliance on expert-driven feature design and sensitivity to imaging variability. In contrast, machine learning-based approaches, particularly deep learning, have demonstrated unparalleled performance in complex image analysis tasks, albeit at the cost of greater data demands and interpretability challenges. As research advances toward 2025, the integration of explainable artificial intelligence (XAI), privacy-preserving technologies like federated learning, and ethical frameworks will likely bridge the gap between technical performance and clinical applicability, shaping the next generation of intelligent medical imaging systems.

3. MACHINE LEARNING FUNDAMENTALS FOR MEDICAL IMAGE ANALYSIS

Machine learning encompasses a diverse set of computational approaches that enable systems to learn patterns from data without explicit programming (Jordan and Mitchell, 2015, Raza and Singh, 2021). In the context of medical image analysis, machine learning algorithms learn to recognize patterns, extract features, and make predictions from examples rather than following predefined rules. This data-driven approach has revolutionized medical image processing, particularly for tasks that are challenging to formalize using traditional algorithmic methods.

3.1. Machine Learning Paradigms in Medical Imaging

Machine learning techniques applied to medical imaging generally fall into three main paradigms: supervised learning, unsupervised learning, and semi-supervised learning (Raza and Singh, 2021). Supervised learning, the most commonly used approach, involves training algorithms on labeled datasets where the desired outputs are known (Sen et al., 2020). In medical imaging, this might include images with expert annotations indicating disease presence, organ boundaries, or tumor locations. The algorithm learns to map input images to appropriate labels, enabling tasks such as classification (e.g., determining whether a lesion is benign or malignant) and segmentation (e.g., delineating organ boundaries). Supervised learning excels when abundant labeled data is available but requires significant effort from medical experts to create reliable ground truth annotations.

Unsupervised learning, conversely, works with unlabeled data, identifying inherent patterns, structures, or relationships without prior knowledge of the correct output (Johnston et al., 2019, Alloghani et al., 2020, Qi et al., 2020). In medical imaging, unsupervised techniques are valuable for dimensionality reduction, feature extraction, anomaly detection, and clustering similar cases. These approaches can reveal novel patterns in imaging data that might not be obvious to human observers and are particularly useful when labeled data is scarce. Clustering algorithms can group similar images or regions, potentially identifying subtypes of diseases with distinct imaging characteristics that may respond differently to treatments.

Semi-supervised learning bridges the gap between supervised and unsupervised approaches by leveraging both labeled and unlabeled data. Given the abundance of unlabeled medical images and the scarcity of expert annotations, semi-supervised methods have gained traction in healthcare applications. These techniques typically train on a small labeled dataset and then extend learning to a larger unlabeled dataset, maximizing the utility of available data while minimizing annotation burden. Active learning, a related approach, strategically selects the most informative cases for expert annotation, optimizing the learning process with minimal labeling effort.

3.2. Traditional Machine Learning Techniques

Before the deep learning revolution, traditional machine learning techniques formed the backbone of medical image analysis (Abdou, 2022, Li et al., 2023). These approaches typically involve a two-stage process: feature extraction followed by classification or regression. Feature extraction transforms raw image data into meaningful representations capturing relevant characteristics such as texture, shape, intensity distributions, and spatial relationships (Salau and Jain, 2019, Kumar et al., 2020). Feature engineering—the process of designing and selecting these features—requires domain expertise and significantly impacts model performance (Zheng and Casari, 2018).

Support Vector Machines (SVMs) have been widely applied in medical image analysis for classification tasks, particularly in cancer detection and staging (Maglogiannis and Zafiropoulos, 2004, Chen et al., 2012). SVMs find optimal hyperplanes separating different classes while maximizing the margin between them, making them effective for high-dimensional data with clear decision boundaries. Random Forests, comprising multiple decision trees that collectively make predictions, offer advantages of interpretability, robustness to outliers, and ability to handle various data types without extensive preprocessing. They have proven effective for many medical image analysis tasks, including organ segmentation and tissue classification.

K-nearest neighbors (KNN) algorithms predict based on similarity to known examples, classifying new cases according to the majority class among their nearest neighbors in the feature space (Mucherino et al., 2009, Cunningham and Delany, 2021). While conceptually simple, KNN can be powerful for medical image analysis when appropriate distance metrics and neighborhood sizes are selected. Principal Component Analysis (PCA) and other dimensionality reduction techniques help manage the high-dimensional nature of medical images by projecting data onto lower-dimensional spaces while preserving essential information, facilitating visualization and subsequent analysis (Maćkiewicz and Ratajczak, 1993).

Table 3 summarizes the traditional machine learning algorithms commonly used in medical image analysis, along with their strengths, limitations, and typical applications.

Table 3. Traditional ML Algorithms in Medical Image Analysis

Algorithm	Description	Strengths	Limitations	Medical Imaging Applications
SVM	Finds optimal hyperplanes to separate classes	- Effective with high dimensional data - Works well with limited training samples - Can capture complex decision boundaries	- Sensitive to parameter selection - May be computationally intensive - Less effective with overlapping classes	- Lesion classification - Disease diagnosis - Tissue characterization
Random Forests	Ensemble of decision trees with majority voting	- Resistant to overfitting - Handles mixed data types - Provides feature importance rankings - Computationally efficient	- Can be memory intensive for large datasets - Less effective with sparse high-dimensional data	- Organ segmentation - Tumor detection - Survival prediction - Multi-class tissue classification
KNN	Classification based on majority vote of k nearest training examples	- Simple implementation - No training phase - Naturally handles multi-class problems	- Computationally expensive during prediction - Sensitive to irrelevant features - Requires feature scaling	- Similar case retrieval - Pathology classification - Simple segmentation tasks
PCA	Dimensionality reduction preserving variance	- Reduces computational complexity - Eliminates redundant features - Enables visualization of high-dimensional data	- Linear assumptions - Sensitive to feature scaling - May lose important information	- Feature extraction - Noise reduction - Dimensionality reduction before classification
GMM	Probabilistic model assuming data generated from mixture of Gaussian distributions	- Flexible clustering approach - Provides probability estimates - Can model complex distributions	- Sensitive to initialization - Requires specifying number of components - Computationally intensive for high dimensions	- Tissue classification - Brain MRI segmentation - Modeling intensity distributions
AdaBoost	Sequential ensemble where each model corrects errors of previous ones	- Reduces bias and variance - Simple to implement - Can incorporate various base learners	- Sensitive to noisy data and outliers - Can overfit with insufficient data	- Lesion detection - Integrating multiple features - Cascade classification systems

3.3. Deep Learning Approaches

Deep learning, a subset of machine learning using artificial neural networks with multiple layers, has transformed medical image analysis by automatically learning hierarchical feature representations directly from data (Suganyadevi et al., 2022, Anwar et al., 2018). Unlike traditional approaches requiring manual feature engineering, deep learning algorithms discover relevant features autonomously, often capturing subtle patterns imperceptible to human experts or conventional algorithms.

CNNs have become the cornerstone of deep learning for medical image analysis (Thakur et al., 2024, Tsuneki, 2022). These specialized neural networks employ convolutional operations that preserve spatial relationships, making them inherently suitable for processing two-dimensional and three-dimensional medical images. The convolutional layers apply filters sliding across the image, detecting features such as edges, textures, and shapes at different scales and abstraction levels. Pooling layers reduce dimensionality while maintaining essential information, and fully connected layers integrate these features for final predictions. This hierarchical architecture enables CNNs to learn complex visual patterns progressively, from simple edges in early layers to organ structures or disease patterns in deeper layers.

As illustrated in Figure 2, the basic architecture of a CNN for medical image classification consists of multiple convolutional layers followed by pooling operations, culminating in fully connected layers that produce the final classification output. This architecture can be adapted for various medical imaging tasks by modifying the network depth, filter configurations, and output layer design.

Figure 2. Basic Architecture of a CNN for Medical Image Classification

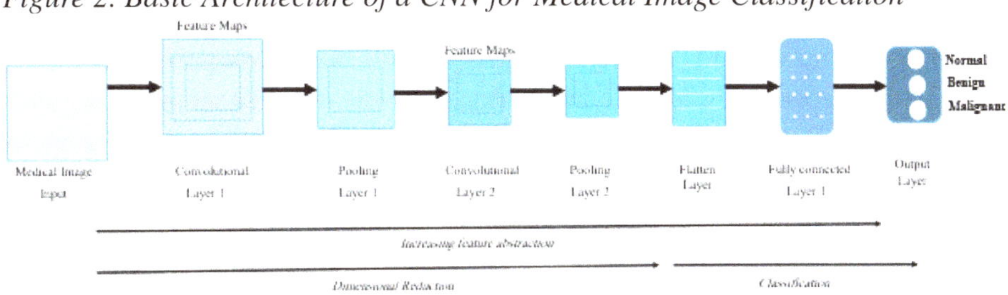

Specialized CNN architecture has been developed for medical image analysis tasks. U-Net, a prominent example, features a symmetric encoder-decoder structure with skip connections, making it particularly effective for segmentation tasks by combining low-level details with high-level contextual information. ResNet and its variants incorporate residual connections that facilitate training of very deep networks by addressing the vanishing gradient problem, enabling more complex feature learning. DenseNet establishes connections between each layer and all subsequent layers, maximizing information flow and feature reuse, which has proven beneficial for medical image classification with limited training data.

Recurrent Neural Networks (RNNs) and their variants, particularly Long Short-Term Memory (LSTM) networks, process sequences of data by maintaining internal memory states (Sherstinsky, 2020). While primarily designed for time-series data, they have been adapted for medical images by treating spatial dimensions as sequences or combining them with CNNs for tasks requiring both spatial and temporal analysis. This hybrid approach has proven valuable for dynamic medical imaging modalities such as cardiac MRI, functional MRI, and ultrasound sequences, where temporal information provides crucial diagnostic insights. In such applications, CNNs typically extract spatial features from individual frames, which are then processed by RNNs to model temporal dependencies across the sequence.

Generative Adversarial Networks (GANs) represent another innovative deep learning architecture comprising two competing networks: a generator that creates synthetic images and a discriminator that distinguishes between real and synthetic images (Showrov et al., 2024). This adversarial training process enables GANs to learn complex data distributions, making them valuable for medical image synthesis, augmentation, domain adaptation, and cross-modality translation. GANs have demonstrated remarkable capabilities in generating realistic medical images from limited datasets, enhancing image quality, and translating between imaging modalities (e.g., MRI to CT), potentially reducing the need for multiple scans and radiation exposure.

Transformers, originally developed for natural language processing, have recently been adapted for medical image analysis, introducing self-attention mechanisms that capture long-range dependencies across image regions (Shamshad et al., 2023). Vision Transformers (ViT) and their medical variants divide images into patches, process them as sequences, and model relationships between distant image regions more effectively than CNNs. This capability has shown promises for whole-slide pathology images, volumetric scans, and other applications where contextual information across the entire image is crucial for accurate interpretation. Table 4 summarizes the deep learning architectures commonly applied to medical image analysis, highlighting their characteristics, strengths, and clinical applications.

Table 4. Deep Learning Architectures for Medical Image Analysis

Architecture	Key Characteristics	Strengths	Medical Imaging Applications
CNN	Specialized for grid-like data with convolutional operations and hierarchical feature learning	- Automated feature extraction - Translation invariance - Parameter sharing - Hierarchical representation learning	- Disease classification - Lesion detection - Feature extraction - General image analysis tasks
U-Net	Encoder-decoder structure with skip connections preserving spatial information	- Precise segmentation with limited training data - Preserves both contextual and detailed information - Effective with heterogeneous structures	- Organ segmentation - Tumor boundary delineation - Cell segmentation in microscopy - Vessel extraction
ResNet	Deep architecture with residual connections enabling training of very deep networks	- Mitigates vanishing gradient problem - Enables training of deeper networks - Improved gradient flow	- Thoracic disease classification - Skin lesion diagnosis - Medical image retrieval - Feature extraction backbone
DenseNet	Each layer connected to all subsequent layers, maximizing feature reuse	- Strong feature propagation - Reduced parameter count - Efficient for limited training data	- Pulmonary nodule classification - Histopathology analysis - Retinal image classification
Recurrent Neural Networks (RNN/ LSTM)	Process sequential data with memory of previous inputs	- Capture temporal dependencies - Handle variable length sequences - Maintain contextual information	- Dynamic imaging analysis (cardiac MRI) - Functional brain imaging - Motion analysis in ultrasound - Medical video processing
GAN	Generator and discriminator networks competing in adversarial training	- Realistic image synthesis - Data augmentation - Domain adaptation - Unsupervised learning	- Cross-modality translation (MRI-CT) - Image enhancement - Data augmentation - Anomaly detection

continued on following page

Table 4. Continued

Architecture	Key Characteristics	Strengths	Medical Imaging Applications
Transformers	Self-attention mechanisms modeling relationships between all input elements	- Capture long-range dependencies - Parallelizable computation - Effective for large-scale images	- Whole-slide pathology analysis - 3D volumetric segmentation - Multi-organ detection - Integrating non-imaging data
Graph Neural Networks (GNN)	Process data represented as graphs with nodes and edges	- Model irregular structures - Incorporate topological information - Integrate multi-scale features	- Brain connectivity analysis - Anatomical structure modeling - Cellular interaction mapping - Vascular network analysis
Auto-encoders	Encoder-decoder architecture for dimensionality reduction and reconstruction	- Unsupervised feature learning - Anomaly detection - Noise reduction	- Image denoising - Feature extraction - Anomaly detection - Dimensionality reduction

3.4 Explainability in Medical AI

The integration of AI into healthcare, particularly in medical image analysis, has shown unprecedented potential in augmenting diagnostic accuracy, automating routine tasks, and enhancing clinical decision-making. However, the "black-box" nature of many ML models, especially deep learning architectures, raises critical concerns regarding transparency, trust, and accountability in clinical applications. This has catalyzed the emergence of Explainable Artificial Intelligence (XAI), a subfield dedicated to making AI decisions more transparent, interpretable, and understandable to human users, particularly clinicians and healthcare stakeholders.

In medical imaging, where diagnostic decisions directly impact patient outcomes, the demand for explainability is not merely a technical preference but a regulatory, ethical, and clinical necessity. This section elaborates on the principal XAI techniques utilized in medical AI, highlighting their mechanisms, applications, and comparative utilities.

Grad-CAM is a visualization technique designed to provide visual explanations for the predictions of CNNs. Grad-CAM generates heatmaps, or saliency maps, that localize the regions within an image that most strongly influence the model's output.

In the context of pulmonary imaging, Grad-CAM has been applied to highlight lung nodules in CT scans, assisting radiologists in identifying suspicious regions corresponding to malignant lesions. The heatmap overlays produced by Grad-CAM provide intuitive visual cues, indicating whether the model is focusing on clinically relevant anatomical structures.

Strengths:

- Highly compatible with CNN-based models.
- Generates class-discriminative localization maps.
- Useful for visual verification by radiologists.

Limitations:

- Relatively coarse localization.
- Dependent on model architecture.

SHAP is a model-agnostic XAI method grounded in cooperative game theory. It assigns each feature an important value (Shapley value) indicating its contribution to a particular prediction.

In breast cancer diagnostics, SHAP has been utilized to elucidate which clinical or imaging-derived features (e.g., tumor size, density, shape, or radiomic signatures) are most influential in the model's classification between benign and malignant lesions. Such interpretability facilitates clinician trust and supports personalized clinical decision-making.

Strengths:

- Global and local interpretability.
- Quantitative assessment of feature contributions.
- Applicability across diverse ML models.

Limitations:

- Computationally intensive for large datasets.
- Interpretation may be abstract in image-based tasks.

LIME operates by locally approximating the complex model with an interpretable surrogate model (e.g., linear regression) around the prediction of interest. This local fidelity provides insights into how slight variations in input features affect the output.

In neuroimaging, LIME has been applied to MRI scans for anomaly detection, offering local interpretability by identifying which specific regions or voxels within an MRI slice contribute most significantly to the detection of tumors or other abnormalities. LIME aids in clarifying individual predictions, especially in ambiguous or borderline cases.

Strengths:

- Local interpretability of individual predictions.
- Model-agnostic.
- Suitable for both tabular and image data.

Limitations:

- Instability in explanations across small perturbations.
- Limited global interpretability.

4. APPLICATIONS FOR MACHINE LEARNING IN MEDICAL IMAGE ANALYSIS

Machine learning has found diverse applications across the medical imaging spectrum, transforming diagnostic processes, treatment planning, and patient monitoring (Pinto-Coelho, 2023, Zhou et al., 2021, Shehab et al., 2022). This section explores key applications in different clinical domains, highlighting how machine learning techniques address specific challenges in medical image analysis.

4.1. Disease Detection and Classification

Disease detection and classification represent primary applications of machine learning in medical imaging, with significant advancements in oncology, cardiology, neurology, and pulmonology (Gonem et al., 2020). In cancer imaging, machine learning algorithms analyze mammograms, chest radiographs, CT scans, and MRI to detect malignancies with increasing accuracy. Deep learning models have demonstrated performance comparable to or exceeding expert radiologists in detecting breast lesions on mammograms, lung nodules on chest CT, and skin cancers from dermatoscopic images. These systems typically employ CNNs trained on large annotated datasets to distinguish malignant from benign findings based on subtle imaging features that might escape human perception.

Cardiovascular disease assessment has benefited substantially from machine learning applications (Litjens et al., 2019). Automated coronary calcium scoring from CT scans helps quantify atherosclerotic burden, while deep learning algorithms analyze cardiac MRI to assess ventricular function, detect myocardial scarring, and identify structural abnormalities. These approaches enable more consistent, quantitative, and efficient cardiac assessment compared to manual methods, potentially leading to earlier intervention and improved patient outcomes.

Neurological disease diagnosis leverages machine learning to detect subtle brain abnormalities on MRI and CT, including early signs of neurodegenerative conditions like Alzheimer's disease, multiple sclerosis, and Parkinson's disease (AbdEl-Latif et al., 2023, Modi et al., 2023, Yadav et al., 2024). Advanced algorithms identify patterns of atrophy, white matter lesions, and connectivity alterations that precede clinical symptoms, potentially enabling earlier intervention. In stroke imaging, machine learning models rapidly identify ischemic regions and assess salvageable tissue, critical for time-sensitive treatment decisions.

Pulmonary disease detection has gained particular prominence with deep learning algorithms that analyze chest radiographs and CT scans to identify tuberculosis, pneumonia, COVID-19, interstitial lung diseases, and chronic obstructive pulmonary disease (Kumar et al., 2024). During the COVID-19 pandemic, machine learning tools were rapidly deployed to triage patients based on imaging findings, demonstrating the adaptability of these approaches to emerging clinical challenges. Figure 3 illustrates the process flow of a typical machine learning system for disease detection, from image acquisition through preprocessing, feature extraction, classification, and decision support.

Figure 3. Machine Learning Process Flow for Pulmonary Disease Detection

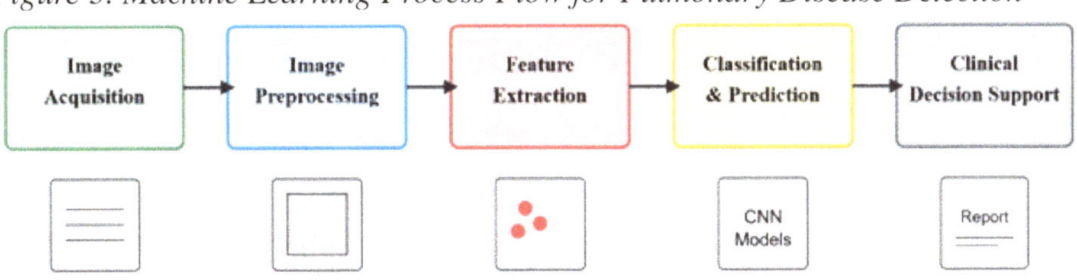

4.2. Image Segmentation

Image segmentation—the process of partitioning images into meaningful regions—represents a fundamental task in medical image analysis with applications across virtually all imaging modalities and anatomical regions (Rogowska, 2009, Withey and Koles, 2008). Machine learning approaches have significantly advanced segmentation capabilities, enabling more precise delineation of organs, tissues, and pathological structures than traditional methods.

Organ and structure segmentation benefits from deep learning techniques that automatically identify anatomical boundaries across imaging modalities (Fu et al., 2021). U-Net and its variants have demonstrated remarkable performance in segmenting complex structures such as cardiac chambers, lungs, liver, kidneys, and brain regions. These segmentations serve as building blocks for subsequent quantitative analysis, treatment planning, and longitudinal monitoring. Automated segmentation reduces inter-observer variability and time-consuming manual delineation, potentially enabling more consistent measurements across clinical settings.

Tumor and lesion segmentation presents particular challenges due to heterogeneous appearances, irregular boundaries, and variable contrast. Deep learning approaches have shown promising results in delineating tumors across multiple anatomical regions, including brain, lung, liver, and breast (Ali et al., 2021, Hussain et al., 2024, Jiang et al., 2023). Precise tumor segmentation facilitates volumetric measurements, characterization of heterogeneity, assessment of treatment response, and radiation therapy planning. The ability to quantify tumor characteristics through segmentation has spawned the field of radiomics, which extracts quantitative features from segmented regions for subsequent analysis.

Vascular segmentation identifies blood vessels in angiographic images, enabling assessment of vascular pathologies such as stenosis, aneurysms, and malformations. Deep learning methods have improved the delineation of complex vascular networks in cerebral, coronary, pulmonary, and peripheral circulation. These segmentations support diagnosis of vascular diseases, surgical planning, and monitoring of interventional procedures. Similar approaches have been applied to airway segmentation for pulmonary assessment and planning bronchoscopic procedures.

Cellular and subcellular segmentation applies machine learning to microscopic images, identifying individual cells, nuclei, and organelles in histopathology, cytology, and microscopy images. These techniques enable automated cell counting, morphological analysis, and classification of cellular abnormalities, supporting pathological diagnosis and biomedical research. The increasing integration of digital pathology with computational analysis has accelerated the development of sophisticated segmentation algorithms for histological images.

Multimodal segmentation leverages complementary information from different imaging techniques to improve accuracy and robustness. Machine learning algorithms can integrate information from structural and functional imaging modalities, such as combining anatomical details from MRI with metabolic information from PET for more precise tumor delineation. These approaches potentially overcome limitations of individual modalities and provide more comprehensive characterization of anatomical structures and pathological processes.

4.3. Image Registration

Image registration, the process of aligning images from different time points, viewpoints, or modalities into a common coordinate system, plays a crucial role in medical image analysis (Rueckert and Schnabel, 2010). Traditional registration methods rely on intensity-based or feature-based optimization, which can be computationally intensive and sensitive to initial conditions. Machine learning approaches have transformed this field by providing faster, more robust, and potentially more accurate registration capabilities.

Deep learning techniques enable direct estimation of transformation parameters between image pairs without iterative optimization. CNNs and other architectures can learn to predict deformation fields or transformation matrices from pairs of images, significantly accelerating the registration process compared to conventional methods. This speed advantage is particularly valuable in time-sensitive clinical scenarios and for processing large imaging datasets.

Multimodal registration aligns images from different imaging techniques (e.g., MRI and CT, PET and MRI) that capture complementary information about anatomical structures and physiological processes. Machine learning approaches address the challenging task of establishing correspondence between images with fundamentally different intensity characteristics by learning common representations across modalities. These techniques facilitate multimodal image fusion for comprehensive diagnosis, surgical planning, and radiation therapy, enabling clinicians to leverage complementary information from different imaging modalities.

Deformable registration accounts for non-rigid anatomical changes between images, such as organ motion, patient positioning differences, or disease progression. Deep learning models can learn complex deformation fields that capture local and global anatomical variations, potentially outperforming traditional parametric transformation models. Applications include motion correction in cardiac and respiratory imaging, longitudinal monitoring of disease progression, and atlas-based segmentation.

Longitudinal registration aligns images from the same patient acquired at different time points, enabling quantitative assessment of disease progression, treatment response, and anatomical changes over time. Machine learning approaches can learn patient-specific patterns of change, potentially providing more sensitive detection of subtle alterations than conventional methods. These techniques are particularly valuable for monitoring neurodegenerative diseases, tumor growth, and treatment effects.

4.4. Computer-Aided Diagnosis

Computer-aided diagnosis (CAD) systems leverage machine learning to assist clinicians in interpreting medical images, potentially improving diagnostic accuracy, efficiency, and consistency (Yeasmin et al., 2024). These systems typically integrate multiple image analysis tasks, including detection, segmentation, feature extraction, and classification, to provide comprehensive diagnostic support.

Screening applications represent a major focus of CAD development, with systems designed to identify abnormalities in screening examinations such as mammography, chest radiography, and colonoscopy. Machine learning algorithms can process large volumes of screening images efficiently, flagging suspicious findings for radiologist review and potentially reducing oversight errors. These approaches address the challenges of screening programs, including high volume, reader fatigue, and subtle early-stage abnormalities.

Diagnostic decision support systems provide quantitative analysis and classification of detected abnormalities, assisting clinicians in characterizing lesions and determining their clinical significance. Modern CAD systems often present probability estimates for different diagnoses rather than binary decisions, supporting a more nuanced interpretation process. These systems may incorporate radiomics features, imaging biomarkers, and clinical information to provide comprehensive diagnostic assessments.

Integrated workflow solutions embed machine learning algorithms within radiology and pathology workflows, prioritizing cases, directing attention to regions of interest, and providing relevant clinical decision support. These workflow-integrated approaches aim to enhance efficiency without disrupting established clinical processes. Advanced systems incorporate natural language processing to extract relevant information from clinical records and generate structured reports based on image analysis.

Second opinion systems offer independent assessments that clinicians can compare with their own interpretations, potentially catching oversights and reducing diagnostic errors. Unlike traditional CAD systems that typically focus on detection, second opinion systems often provide comprehensive interpretations more similar to those of radiologists. These systems may be particularly valuable in settings with limited specialist access or for complex cases warranting additional review. Figure 4 illustrates the key components and workflow of a modern computer-aided diagnosis system, highlighting the integration of multiple machine learning techniques to support clinical decision-making.

Figure 4. Modern Computer-Aided Diagnosis System Workflow

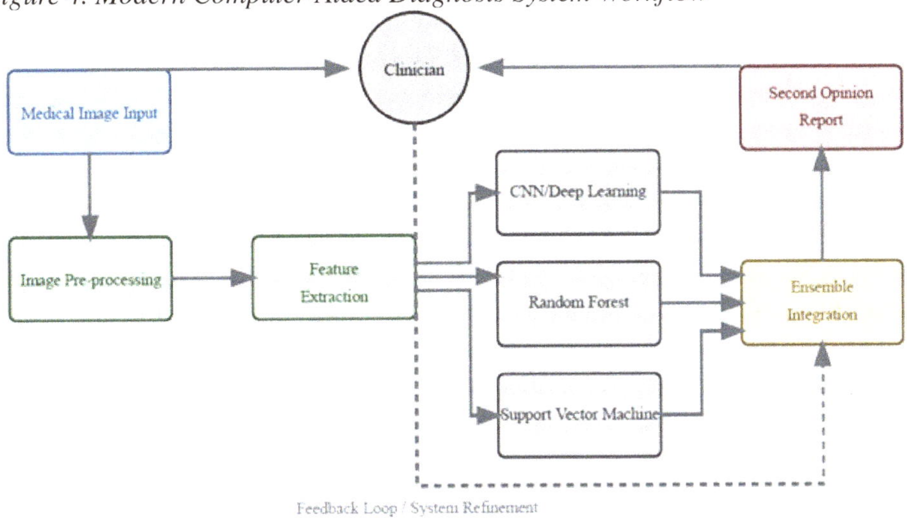

4.5. Treatment Planning and Monitoring

Machine learning substantially impacts treatment planning and monitoring across various medical specialties, optimizing therapeutic strategies and assessing treatment effectiveness. These applications leverage precise segmentation, predictive modeling, and longitudinal analysis capabilities to personalize interventions and track patient responses.

Radiation therapy planning benefits from machine learning-based segmentation of tumors and critical structures, enabling more precise targeting of cancerous tissues while sparing healthy organs. Automated segmentation reduces the time-consuming manual contouring process and potentially improves consistency across treatment centers. Advanced algorithms can predict optimal dose distributions, suggest beam arrangements, and even estimate treatment outcomes based on imaging and clinical characteristics. These approaches facilitate adaptive radiotherapy, where treatment plans are modified based on anatomical changes detected during the course of therapy.

Surgical planning applications incorporate machine learning to segment anatomical structures, identify critical neurovascular structures, and simulate procedural outcomes. These tools support preoperative decision-making by providing three-dimensional visualizations of patient-specific anatomy and pathology. In neurosurgery, machine learning assists in mapping functional brain regions and white matter tracts to minimize postoperative deficits. Similar approaches guide orthopedic procedures by analyzing biomechanical factors and predicting optimal implant positioning.

Interventional procedure guidance leverages real-time image analysis to assist minimally invasive procedures such as catheter-based interventions, needle biopsies, and ablations. Machine learning algorithms can track instruments, register preoperative plans with intraoperative imaging, and highlight anatomical landmarks, potentially improving procedural accuracy and reducing complications. These applications increasingly incorporate augmented reality displays that overlay segmentation and navigational information onto the interventional field of view.

Treatment response assessment employs longitudinal image analysis to evaluate therapeutic effectiveness, comparing pre- and post-treatment imaging to quantify changes in tumor size, metabolic activity, or other relevant biomarkers. Machine learning approaches extract quantitative metrics from sequential images, potentially detecting subtle response patterns that might escape visual assessment. These techniques support objective evaluation according to standardized criteria such as RECIST (Response Evaluation Criteria in Solid Tumors) while potentially identifying novel imaging biomarkers predictive of long-term outcomes (Fournier et al., 2022).

Outcome prediction models integrate imaging features with clinical, genetic, and laboratory data to forecast treatment responses and long-term prognosis. These predictive models assist in treatment selection by identifying patients likely to benefit from specific interventions or requiring more aggressive approaches. By incorporating radiomics features and deep learning-derived imaging biomarkers, these models can potentially capture subtle prognostic information beyond conventional imaging assessments, supporting personalized treatment decisions. Table 5 summarizes machine learning applications in treatment planning and monitoring across different therapeutic modalities, highlighting the impact on clinical workflow and patient outcomes.

Table 5. Machine Learning Applications in Treatment Planning and Monitoring

Therapeutic Modality	Machine Learning Applications	Impact on Clinical Workflow	Patient Outcome Benefits
Radiation Therapy	- Automated tumor and critical structure segmentation - Dose distribution optimization - Beam arrangement prediction - Treatment outcome estimation - Adaptive plan generation	- 60-80% reduction in contouring time - Standardized planning across institutions - Faster treatment planning process - Real-time plan adaptation capabilities	- Improved target coverage - Reduced damage to healthy tissues - Decreased treatment-related toxicity - Potential for dose escalation to resistant regions
Surgical Planning	- 3D anatomical reconstruction - Critical structure identification - Surgical approach optimization - Procedural outcome simulation - Functional tissue mapping	- Enhanced preoperative visualization - Improved surgical team communication - Reduced planning time for complex cases - Automated surgical risk assessment	- Decreased operative time - Reduced complication rates - Preservation of critical structures - Improved functional outcomes
Interventional Procedures	- Real-time instrument tracking - Intraoperative registration - Anatomical landmark detection - Augmented reality guidance - Procedural risk prediction	- Simplified navigation in complex anatomy - Reduced reliance on contrast agents - Decreased fluoroscopy time - Enhanced visualization during procedures	- Increased procedural accuracy - Reduced radiation exposure - Lower complication rates - Shorter recovery times
Treatment Response Assessment	- Automated RECIST measurements - Early response pattern detection - Subtle change quantification - Novel biomarker extraction - Pseudoprogression vs. true progression differentiation	- Consistent longitudinal measurements - Reduced inter-observer variability - Automated report generation - Facilitated multi-timepoint comparison	- Earlier treatment modification for non-responders - Reduced unnecessary treatment changes - More accurate prognostication - Objective response documentation
Drug Therapy	- Dose optimization algorithms - Adverse effect prediction - Drug resistance modeling - Therapeutic response forecasting - Patient-specific regimen design	- Data-driven dosing decisions - Proactive toxicity management - Streamlined selection - Predictive monitoring alerts	- Improved therapeutic efficacy - Reduced adverse events - Optimized medication adherence - Personalized treatment approaches

continued on following page

Table 5. Continued

Therapeutic Modality	Machine Learning Applications	Impact on Clinical Workflow	Patient Outcome Benefits
Rehabilitation	- Movement pattern analysis - Functional improvement prediction - Exercise optimization - Recovery trajectory modeling - Compensatory mechanism detection	- Objective progress assessment - Automated exercise prescription - Remote monitoring capabilities - Personalized rehabilitation plans	- Accelerated functional recovery - Improved adherence to therapy - Enhanced quality of life - Better long-term functional outcomes
Palliative Care	- Symptom trajectory prediction - Intervention timing optimization - Quality of life assessment - Personalized comfort measures - End-of-life planning support	- Proactive symptom management - Enhanced resource allocation - Improved interdisciplinary communication - Timely palliative intervention	- Better symptom control - Reduced emergency department visits - Enhanced patient comfort - Improved quality of remaining life

4.6. Image Quality Enhancement and Reconstruction

Machine learning approaches have revolutionized medical image enhancement and reconstruction, addressing fundamental limitations in image acquisition while potentially improving diagnostic accuracy and reducing radiation exposure. These techniques leverage the ability of deep learning models to learn complex mappings between degraded and high-quality images.

Noise reduction and artifact removal employs deep learning to mitigate image degradation from various sources, including hardware limitations, motion, and acquisition protocols. Convolutional neural networks and generative models have demonstrated remarkable capability in distinguishing signal from noise without blurring important anatomical details. These approaches have been successfully applied to denoise low-dose CT scans, reduce MRI artifacts, and enhance ultrasound images. By enabling diagnostic-quality imaging with reduced radiation or shortened acquisition times, these techniques potentially improve patient safety and throughput.

Super-resolution techniques enhance spatial resolution beyond the limitations of imaging hardware, generating high-resolution images from low-resolution inputs. Deep learning models learn to predict fine details based on contextual information and prior knowledge of anatomical structures. Applications include enhancing resolution in MRI, improving visualization of small structures in ultrasound, and refining details in nuclear medicine imaging. These approaches potentially extend the capabilities of existing imaging equipment without hardware upgrades.

Image reconstruction algorithms have been transformed by deep learning, particularly in modalities that require computational reconstruction from raw data, such as CT, MRI, and PET. Machine learning approaches can reconstruct higher quality images from limited or noisy acquisition data compared to conventional analytical methods. Significant applications include accelerated MRI reconstruction from undersampled k-space data, enabling faster scan times without compromising quality, and low-dose CT reconstruction that maintains diagnostic accuracy with reduced radiation exposure. These techniques increasingly incorporate physics-based models within deep learning frameworks, ensuring that reconstructions adhere to physical imaging principles.

Multimodal image synthesis generates images with characteristics of one modality based on information from another, potentially overcoming limitations in image acquisition or availability. For example, deep learning models can synthesize CT-like images from MRI data, PET-like functional information from structural imaging, or specialized MRI sequences from standard acquisitions. These approaches potentially reduce the need for multiple scans, decrease radiation exposure, and provide comprehensive information in situations where certain imaging modalities are contraindicated or unavailable.

4.7. Data Integration and Multimodal Analysis

The integration of multiple data sources with medical images represents an emerging frontier in machine learning applications, enabling more comprehensive analysis and personalized medicine approaches (Li et al., 2023, Nananukul et al., 2024). These techniques combine imaging findings with clinical, pathological, genomic, and other patient data to provide richer context for diagnosis and treatment planning.

Radiogenomic analysis correlates imaging features with genomic characteristics, potentially enabling non-invasive prediction of genetic mutations, expression patterns, and molecular subtypes. Machine learning algorithms identify relationships between imaging phenotypes and underlying molecular profiles, supporting precision oncology applications where treatment decisions increasingly depend on genetic information. These approaches potentially reduce the need for invasive biopsies by inferring genomic information from standard imaging studies, particularly valuable when tissue sampling is risky or challenging.

Integrated diagnostic systems combine imaging data with electronic health records, laboratory results, pathology findings, and clinical metrics to provide comprehensive diagnostic assessments. Machine learning models trained on these diverse data sources can potentially identify complex patterns beyond those apparent in any single data type. These systems increasingly incorporate natural language processing to extract relevant information from unstructured clinical notes and reports, creating structured representations for algorithmic analysis.

Longitudinal patient modeling tracks changes across multiple data streams over time, potentially identifying early signs of disease progression or treatment response before conventional clinical assessments. Machine learning approaches can model complex temporal patterns across imaging and non-imaging data, accounting for individual patient variability. These models support personalized risk stratification and treatment adaptation based on individual disease trajectories rather than population averages.

Digital twin concepts apply comprehensive computational modeling to create patient-specific representations that integrate multiple data sources and can predict individual responses to interventions. These approaches increasingly incorporate medical imaging as a core component, providing anatomical and functional context for physiological simulations. Machine learning techniques enable model personalization based on patient-specific data, potentially supporting virtual treatment planning and outcome prediction.

4.8. Future Directions and Challenges

As machine learning continues to transform medical image analysis, several emerging directions and persistent challenges will shape future developments in this rapidly evolving field. Addressing these challenges will be crucial for translating promising research into clinical impact.

Federated learning and distributed analysis approaches enable collaborative model training across institutions without sharing sensitive patient data, potentially addressing privacy concerns while leveraging diverse datasets. These techniques allow hospitals and research centers to contribute to model development while keeping patient images within their secure environments. Such approaches may accelerate the development of robust algorithms by incorporating data diversity across geographic regions, patient demographics, and institutional practices.

Explainable AI for medical imaging addresses the "black box" nature of many deep learning systems by developing techniques that make algorithmic decisions more transparent and interpretable. Attention mechanisms, feature visualization, and interpretable architectures help clinicians understand the basis for machine learning predictions, potentially increasing trust and facilitating regulatory approval. These approaches balance performance with explainability, ensuring that clinicians can appropriately integrate algorithmic outputs into clinical decision-making.

Generalizability across diverse populations and imaging protocols remains a significant challenge for machine learning models. Algorithms developed and validated in specific settings often show performance degradation when applied to different patient demographics, equipment manufacturers, or acquisition parameters. Techniques to improve robustness include diverse training data, domain adaptation methods, and adversarial training approaches that explicitly account for variations in imaging characteristics. Addressing these generalizability challenges is essential for equitable implementation of machine learning tools across healthcare settings.

Integration into clinical workflows presents practical challenges beyond technical performance, including user interface design, integration with existing systems, and alignment with clinical processes. Successful implementation requires close collaboration between technical developers and clinical stakeholders to ensure that machine learning tools enhance rather than disrupt established workflows. Considerations include seamless PACS (Picture Archiving and Communication System) integration, appropriate presentation of algorithmic outputs, and efficient mechanisms for clinician review and feedback.

Regulatory considerations and validation standards continue to evolve as machine learning applications proliferate in medical imaging. Regulatory bodies are developing frameworks to evaluate "adaptive" algorithms that may change over time, ensure appropriate performance across diverse patient populations, and monitor deployed systems for unexpected behaviors. Standardized evaluation metrics, benchmark datasets, and reporting guidelines aim to facilitate consistent assessment of algorithm performance and support regulatory review processes.

5. PERFORMANCE EVALUATION AND VALIDATION IN MEDICAL IMAGE ANALYSIS

Performance evaluation and validation constitute essential components in the development, assessment, and clinical translation of machine learning systems for medical image analysis. Rigorous evaluation methodologies are necessary to establish algorithmic efficacy, compare approaches, ensure generalizability, and ultimately gain regulatory approval and clinical acceptance. This section explores key metrics, methodologies, and challenges in evaluating machine learning performance within the healthcare imaging domain.

5.1. Evaluation Metrics and Methodologies

Evaluation metrics for machine learning in medical imaging vary according to the specific task being performed. For classification tasks such as disease detection or lesion characterization, standard metrics include sensitivity (recall), specificity, accuracy, precision, F1-score, and area under the receiver operating characteristic curve (AUC-ROC) (Naidu et al., 2023). These metrics capture different aspects of classification performance, with sensitivity measuring the ability to correctly identify positive cases and specificity assessing correct identification of negative cases. The AUC-ROC provides a comprehensive performance assessment across different decision thresholds, making it particularly valuable for probabilistic outputs that characterize many deep learning systems.

For segmentation tasks, evaluation metrics focus on spatial overlap and distance between predicted and ground truth delineations. The Dice similarity coefficient (DSC) and Jaccard index quantify volumetric overlap between segmentations, while Hausdorff distance and average surface distance measure boundary agreement. These metrics may be applied to entire structures or specific regions of interest, depending on clinical priorities. Volumetric measures such as absolute and percentage volume difference complement overlap metrics by capturing systematic under- or over-segmentation that might have clinical significance despite reasonable spatial overlap.

Regression tasks, such as quantifying cardiac ejection fraction or tumor volume, employ metrics including mean absolute error (MAE), root mean squared error (RMSE), and correlation coefficients between predicted and reference values. These metrics assess the precision and accuracy of continuous predictions, with different metrics emphasizing different aspects of performance. For example, RMSE penalizes large errors more heavily than MAE, which may be appropriate when large deviations have disproportionate clinical consequences.

Beyond standard quantitative metrics, clinical validation increasingly incorporates task-based assessment, where algorithm performance is evaluated based on its impact on clinical decision-making rather than purely technical criteria. This approach recognizes that technical performance may not translate directly to clinical utility. For instance, an algorithm with slightly lower segmentation accuracy might be clinically preferable if it consistently includes clinically significant regions that another algorithm occasionally misses. Table 6 summarizes key evaluation metrics for common medical image analysis tasks, highlighting their mathematical definitions, interpretations, and clinical relevance.

Table 6. Evaluation Metrics for Medical Image Analysis Tasks

Metric	Task Type	Mathematical Definition	Interpretation	Clinical Relevance
Sensitivity (Recall)	Classification	TP/(TP+FN)	Proportion of actual positives correctly identified	Critical for screening applications where missing cases has serious consequences
Specificity	Classification	TN/(TN+FP)	Proportion of actual negatives correctly identified	Important for avoiding unnecessary follow-up procedures
Precision	Classification	TP/(TP+FP)	Proportion of positive predictions that are correct	Relevant when false positives have significant impact
F1-Score	Classification	2×(Precision×Recall)/ (Precision+Recall)	Harmonic mean of precision and recall	Balanced measure when both false positives and negatives are important
AUC-ROC	Classification	Area under ROC curve	Classification performance across thresholds	Comprehensive assessment of discriminative ability
Hausdorff Distance	Segmentation	max(h(A,B), h(B,A)) where h(A,B)=max(min(d(a,b)))	Maximum surface distance between segmentations	Sensitive to outliers, captures worst-case boundary disagreement
Mean Absolute Error	Regression	(1/n)×Σ|y_i-ŷ_i|	Average absolute difference between prediction and truth	Intuitive measure of prediction accuracy
Correlation Coefficient	Regression	cov(Y,Ŷ)/σ_Y×σ_Ŷ	Linear relationship between predicted and true values	Indicates prediction consistency across value range
Mean Opinion Score	Image Quality	Average of expert quality ratings	Subjective quality assessment	Captures clinically relevant image quality aspects

5.2. Validation Strategies

Robust validation strategies are essential for ensuring that machine learning models for medical image analysis perform reliably across diverse clinical scenarios (Parmar et al., 2018). Cross-validation techniques such as k-fold cross-validation partition available data into training and validation sets multiple times, providing performance estimates with reduced variance compared to a single train-test split (Allgaier and Pryss, 2024). These approaches help assess model stability and identify potential overfitting, particularly valuable when working with limited datasets. Stratified cross-validation ensures balanced representation of important subgroups (e.g., disease stages or demographic categories) across partitions, addressing potential biases in performance estimation.

External validation on independent datasets represents the gold standard for assessing generalizability, evaluating model performance on data from different institutions, patient populations, or acquisition protocols than those used during development. This validation approach identifies potential overfitting to specific dataset characteristics and assesses real-world applicability. Multi-center validation studies increasingly incorporate data heterogeneity as a deliberate design element, testing model robustness across diverse clinical settings.

Temporal validation assesses model performance on newer data collected after model development, simulating the natural evolution of clinical practice over time. This approach captures potential dataset shift due to changes in imaging protocols, patient demographics, or disease patterns. Temporal validation is particularly relevant for applications involving longitudinal monitoring or deployment in evolving clinical environments.

Reader studies compare algorithm performance against human experts, typically having multiple clinicians interpret the same cases with and without algorithmic assistance. These studies measure both standalone algorithm performance and its impact on human diagnostic accuracy, efficiency, and consistency. Well-designed reader studies employ appropriate statistical methods to account for case difficulty and inter-reader variability, providing insights into how algorithms might complement human expertise in clinical practice.

5.3. Challenges in Evaluation

Despite advances in evaluation methodologies, significant challenges persist in assessing machine learning systems for medical image analysis. Reference standard limitations represent a fundamental challenge, as ground truth annotations often derive from expert interpretations subject to inter- and intra-observer variability rather than definitive ground truth. This variability creates an inherent ceiling on measurable algorithm performance and complicates performance comparisons. Approaches to address this challenge include consensus annotations from multiple experts, adjudication of disagreements, and statistical methods that account for reference uncertainty.

Dataset bias and representativeness issues arise when training or evaluation data fail to capture the diversity encountered in clinical practice. Models evaluated on homogeneous datasets may show artificially high performance that deteriorates when applied to underrepresented subpopulations or imaging conditions. Addressing this challenge requires diverse datasets spanning different demographics, disease presentations, imaging protocols, and equipment manufacturers, along with explicit reporting of dataset characteristics and limitations.

Performance metrics may not always align with clinical utility, as standard technical measures might emphasize aspects with limited clinical relevance while overlooking critical factors for practical implementation. For example, perfect segmentation of clinically insignificant regions may inflate overall performance metrics while providing minimal clinical value. Addressing this challenge requires close collaboration between technical developers and clinical stakeholders to define clinically meaningful evaluation criteria and validation approaches.

Implementation-specific factors such as computational efficiency, inference time, and hardware requirements significantly impact clinical utility yet often receive less attention than accuracy metrics in academic research. Comprehensive evaluation should consider these practical aspects, particularly for applications requiring real-time processing or deployment on resource-constrained systems.

6. CHALLENGES AND LIMITATIONS IN MACHINE LEARNING FOR MEDICAL IMAGING

Despite remarkable progress, machine learning applications in medical imaging face substantial challenges that must be addressed to realize their full potential in clinical practice. These challenges span technical, methodological, clinical, and ethical dimensions, requiring interdisciplinary approaches for effective resolution.

6.1. Data Availability and Quality Challenges

Limited labeled data represents a persistent challenge in medical imaging applications, as expert annotations are time-consuming, expensive, and subject to variability. While natural image datasets often contain millions of labeled examples, medical imaging datasets typically include hundreds or thousands of cases, limiting the complexity of learnable patterns. This challenge particularly affects rare conditions or specialized applications where collecting sufficient examples proves difficult. Approaches to address data limitations include transfer learning from models pretrained on larger datasets, data augmentation techniques that artificially expand training data through transformations, and self-supervised learning methods that leverage unlabeled data.

Data quality and standardization issues stem from variations in imaging protocols, equipment manufacturers, and institutional practices that introduce heterogeneity beyond clinically relevant differences. Inconsistent image resolution, contrast, noise characteristics, and artifacts complicate model development and generalization. Preprocessing and normalization techniques can partially mitigate these issues, but fundamental variations often remain. Standardization initiatives across healthcare institutions aim to harmonize acquisition protocols and quality control procedures, potentially reducing unnecessary variability while preserving clinically meaningful differences.

Class imbalance characterizes many medical imaging applications, where normal findings substantially outnumber abnormal ones or certain disease subtypes appear rarely. This imbalance can bias machine learning models toward majority classes, compromising performance on clinically important minority classes. Techniques to address class imbalance include weighted loss functions that emphasize under-represented classes, oversampling of minority cases, undersampling of majority cases, and ensemble methods that combine multiple balanced submodels. These approaches help ensure adequate learning of rare but clinically significant patterns.

Privacy and ethical considerations significantly impact data availability for machine learning in healthcare. Patient privacy regulations and ethical frameworks rightly restrict data sharing and usage, complicating multi-institutional collaboration and public dataset creation. Privacy-preserving techniques such as federated learning, differential privacy, and synthetic data generation offer promising avenues for enabling collaborative model development while protecting patient confidentiality. These approaches allow institutions to contribute to model training without directly sharing sensitive patient data, potentially unlocking larger and more diverse datasets for algorithm development.

6.2. Technical and Methodological Challenges

Model interpretability remains a significant challenge for many deep learning architectures applied to medical imaging. Complex models with millions of parameters often function as "black boxes," providing predictions without clear explanations of their reasoning. This opacity raises concerns among clinicians and regulators regarding trust, safety, and potential biases. Explainable AI techniques such as attention maps, feature visualization, saliency methods, and interpretable architectures aim to provide insights into model decision processes. These approaches help clinicians understand and verify model reasoning, potentially increasing trust and facilitating integration into clinical workflows.

Generalizability across different patient populations, imaging protocols, and clinical settings presents substantial challenges for machine learning models. Models often demonstrate performance degradation when applied to data differing from their training distribution, limiting clinical utility across diverse healthcare environments. Domain adaptation techniques, transfer learning, adversarial robustness methods, and explicit modeling of population variability aim to improve cross-domain performance. Developing models that maintain consistent performance across diverse clinical scenarios remains an active research area essential for equitable healthcare implementation.

Uncertainty quantification enables models to express confidence in their predictions, a critical capability for medical applications where knowing when to defer to human judgment can prevent harmful errors. Methods for uncertainty estimation include Bayesian neural networks, ensemble approaches, dropout sampling, and direct uncertainty prediction. Properly calibrated uncertainty estimates can guide clinical decision-making, determining when algorithmic suggestions should be trusted or questioned. Integrating uncertainty information into clinical interfaces represents an emerging direction for responsible AI deployment in healthcare.

Computational efficiency considerations become increasingly important as models grow more complex and datasets larger. State-of-the-art deep learning architectures often require substantial computational resources for training and inference, potentially limiting deployment in resource-constrained settings. Techniques including model compression, knowledge distillation, quantization, and hardware-optimized implementations aim to reduce computational requirements while maintaining performance. Balancing model sophistication with practical deployment constraints remains an important consideration for clinical translation.

6.3. Clinical Integration Challenges

Workflow integration represents a critical yet often underappreciated challenge in deploying machine learning systems for medical imaging. Even technically excellent algorithms may fail to achieve clinical impact if they disrupt established workflows or increase clinician burden. Successful integration requires careful human factors engineering, seamless connections with existing systems (PACS, electronic health records), appropriate alert mechanisms, and interfaces aligned with clinical decision processes. Co-designing systems with end-users helps ensure that machine learning tools complement rather than complicate clinical workflows.

Clinical validation beyond technical performance metrics requires demonstrating improved patient outcomes, clinical decision quality, or healthcare efficiency. This validation involves prospective clinical trials evaluating machine learning systems in real-world settings, measuring impacts on diagnostic accuracy, treatment decisions, workflow efficiency, and ultimately patient outcomes. Such studies typically require longer timeframes and greater resources than technical evaluations but provide essential evidence for clinical adoption and regulatory approval.

Regulatory considerations vary across jurisdictions but generally focus on safety, efficacy, and risk management for machine learning applications in medical imaging. Regulatory frameworks continue to evolve as agencies develop approaches for evaluating "adaptive" algorithms that may change over time through continued learning. Key considerations include performance validation across diverse populations, monitoring deployed systems for unexpected behaviors, managing algorithm updates, and maintaining documentation throughout the product lifecycle. Navigate these requirements necessitates interdisciplinary collaboration among technical developers, clinical experts, and regulatory specialists.

Liability and responsibility questions emerge as machine learning systems increasingly influence clinical decisions. Understanding who bears responsibility for errors—algorithm developers, healthcare providers, institutions, or some combination—remains an evolving legal and ethical question. Clear communication of algorithm capabilities and limitations, appropriate training for clinical users, and explicit policies regarding oversight and intervention help address these concerns. As machine learning systems become more autonomous, frameworks for shared responsibility between human and artificial intelligence will require further development.

The Food and Drug Administration (FDA) has established comprehensive regulatory guidelines specifically targeting Software as a Medical Device (SaMD), which refers to software intended for medical purposes without being part of a hardware medical device. The foundational framework for SaMD regulation is derived from the International Medical Device Regulators Forum (IMDRF), which emphasizes a risk-based approach to software evaluation. Key elements of the FDA SaMD regulatory process include:

- Risk Categorization: Based on the intended use and potential impact on patient safety if the software were to malfunction.
- Clinical Evaluation: Demonstration of clinical validity and performance through well-designed clinical studies or retrospective analysis of real-world data.

- Software Verification and Validation: Ensures software reliability, performance consistency, and cybersecurity.
- Human Factors and Usability Engineering: Assessing the user interface to minimize operational errors in clinical settings.
- Post-Market Surveillance: Continuous monitoring and reporting of software performance post-approval, including updates and error corrections.

The FDA also introduced a Digital Health Software Precertification (Pre-Cert) Program aimed at accelerating the approval process for SaMD products while maintaining high safety and efficacy standards.

In parallel, the European Union (EU) mandates that SaMD products must obtain the CE (Conformité Européenne) marking to be legally marketed within its member states. Under the new Medical Device Regulation (MDR 2017/745), software products are classified based on intended purpose and associated risks. Key requirements for CE marking include:

- Technical Documentation: Comprehensive documentation detailing software design, development processes, risk management, and performance evaluation.
- Clinical Evaluation Report (CER): Analysis demonstrating conformity with clinical performance expectations, often requiring evidence from clinical trials or post-market data.
- Quality Management Systems (QMS): Compliance with standards such as ISO 13485 for medical devices.
- Notified Body Assessment: Independent evaluation of technical and clinical evidence by a certified Notified Body.
- Post-Market Clinical Follow-up (PMCF): Ongoing assessment of the device's safety and performance in real-world settings.

A landmark example of successful clinical integration of an AI-based medical image analysis tool is the IDx-DR system, the first autonomous AI diagnostic system approved by the FDA in 2018 for the detection of diabetic retinopathy. IDx-DR exemplifies the regulatory pathway and validation rigor required for clinical adoption of SaMD technologies. DA Approval Process of IDx-DR:

- Risk Categorization: Classified as a Class II medical device.
- Clinical Validation: A pivotal multi-center clinical trial involving 900 adult patients demonstrated that IDx-DR achieved a sensitivity of 87.4% and specificity of 89.5% for detecting more-than-mild diabetic retinopathy.

- Autonomous Operation: Approved for use without clinician interpretation, representing a paradigm shift towards autonomous AI diagnostics.
- Human Factors Evaluation: Assessed for usability in primary care settings by non-specialist staff.
- Post-Market Requirements: Ongoing monitoring to ensure continued performance in diverse populations.

IDx-DR has also received CE marking approval in the European Union, validating its compliance with EU medical device regulations, including software safety, performance, and cybersecurity considerations.

While regulatory approval is a critical milestone, broader clinical integration challenges persist, including:

- Data Generalizability: Ensuring model robustness across diverse demographics, imaging devices, and clinical settings.
- Interoperability: Seamless integration with existing Electronic Health Record (EHR) systems and Picture Archiving and Communication Systems (PACS).
- Ethical and Legal Considerations: Addressing liability in cases of AI misdiagnosis and ensuring transparency in decision-making (explainable AI).
- Clinician Acceptance: Overcoming skepticism and building trust among healthcare providers through explainability, user-friendly interfaces, and demonstrable clinical benefit.
- Continuous Learning Systems: Balancing regulatory requirements with the need for AI systems to update and adapt over time without compromising patient safety.

6.4. Ethical and Social Considerations

Algorithmic bias presents a significant concern for machine learning in healthcare, potentially perpetuating or amplifying existing healthcare disparities if training data reflect historical inequities in care access or quality. Models trained on demographically skewed datasets may perform poorly on underrepresented populations or incorporate biased patterns in their predictions. Mitigation approaches include diverse and representative training data, explicit fairness constraints during model development, performance disaggregation across demographic groups, and ongoing monitoring for disparate impact. Ensuring that machine learning systems promote health equity rather than undermine it remains an essential consideration.

Transparency and accountability mechanisms are necessary to build trust among patients, providers, and regulators. These mechanisms include clear documentation of model development processes, explicit disclosure of limitations and intended use cases, appropriate performance metrics across population subgroups, and systems for monitoring and addressing issues after deployment. Creating institutional structures for algorithmic governance and oversight helps ensure responsible development and deployment practices.

Access and equity concerns arise when advanced machine learning technologies deploy unevenly across healthcare settings, potentially widening disparities between well-resourced and under-resourced institutions. Developing algorithms that function effectively with limited computational resources, creating sustainable implementation and support models for diverse healthcare environments, and prioritizing high-impact applications for underserved populations can help address these concerns. International collaborations and open-source initiatives can further promote equitable access to algorithmic innovations.

Patient autonomy and informed consent take on new dimensions with AI-assisted healthcare. Patients have the right to understand when algorithms influence their care and the limitations of these systems. Developing appropriate disclosure frameworks, accessible explanations of algorithmic processes, and mechanisms for patients to opt out when desired supports autonomous decision-making. Patient involvement in algorithm development and governance further ensures that these technologies align with diverse patient values and preferences.

7. EMERGING TRENDS AND FUTURE DIRECTIONS

The field of machine learning for medical image analysis continues to evolve rapidly, with emerging approaches addressing existing limitations and opening new application areas. This section explores key trends likely to shape future developments in this dynamic domain.

7.1. Advanced Learning Paradigms

Self-supervised learning techniques leverage unlabeled data by creating artificial supervisory signals from the data itself, reducing dependence on expert annotations. These approaches pre-train models on tasks such as reconstructing masked image regions, predicting image transformations, or contrasting different views of the same image. The resulting representations capture meaningful anatomical and pathological patterns that transfer effectively to downstream tasks with limited labeled data. Self-supervised pre-training increasingly serves as a foundation for medical image analysis, particularly valuable given the abundance of unlabeled medical images compared to expert annotations.

Few-shot and zero-shot learning methods aim to recognize new conditions or anatomical structures with minimal or no specific training examples. These approaches leverage similarities between known and novel categories, meta-learning strategies that learn how to learn efficiently, or large-scale pre-training that captures generalizable visual concepts. Such techniques could enable more flexible diagnostic systems that adapt to rare conditions or emerging diseases without requiring extensive retraining with new labeled datasets.

Continual learning addresses the challenge of model adaptation over time without catastrophic forgetting of previously learned knowledge. As imaging protocols evolve, new disease patterns emerge, and clinical practices change, machine learning systems must adapt without requiring complete retraining or losing valuable learned representations. Approaches including parameter regularization, memory replay, and architecture adaptation enable models to incorporate new knowledge while preserving previous capabilities. These techniques support sustainable deployment of machine learning systems in dynamic clinical environments.

Reinforcement learning (RL) frameworks train agents through interaction with environments, optimizing sequences of actions to maximize cumulative rewards. Medical applications of RL include optimizing image acquisition parameters, automated view planning, adaptive radiation therapy, and sequential decision support for treatment selection. Unlike supervised learning, RL can directly optimize for long-term clinical outcomes rather than proxy metrics, potentially addressing the gap between technical performance and clinical utility. Combining RL with realistic medical simulators enables training without risking patient safety during the learning process.

7.2. Multimodal and Integrated Approaches

Multimodal fusion techniques integrate information from complementary imaging modalities, capturing structural, functional, and molecular aspects of pathology that may not be apparent in any single modality. Advanced deep learning architectures including cross-attention mechanisms, multistream networks, and graph-based approaches enable effective integration of heterogeneous imaging data while preserving modality-specific characteristics. These fusion approaches increasingly incorporate optimal alignment and registration as integrated components rather than separate preprocessing steps. Clinical applications span neurology (combining structural MRI, functional MRI, and diffusion tensor imaging), oncology (integrating CT, MRI, and

PET), and cardiology (fusing anatomical and functional cardiac imaging), enabling more comprehensive disease characterization and precise treatment planning.

Image-omics integration combines imaging data with genomic, proteomic, or metabolomic information to establish relationships between imaging phenotypes and molecular characteristics. Machine learning approaches identify correlations between imaging features and underlying biological processes, potentially enabling non-invasive molecular profiling through imaging. Multimodal architectures that process imaging and molecular data in parallel streams before integration at higher levels have demonstrated promising results in cancer subtyping, treatment response prediction, and prognostic assessment. This emerging field bridges radiology and pathology with molecular medicine, supporting precision healthcare approaches that tailor interventions to individual patient characteristics.

Clinical context integration incorporates relevant non-imaging patient data—including demographics, laboratory values, medical history, and treatment information—into image analysis pipelines. These approaches recognize that images exist within broader clinical contexts that influence their interpretation and significance. Architectures for clinical context integration range from early fusion (combining clinical data with image features before analysis) to late fusion (independent processing with integrated decision-making) and interactive approaches that dynamically weight information sources based on their relevance. Such systems potentially overcome limitations of image-only analysis by incorporating critical contextual factors that affect diagnosis and treatment decisions.

Digital twins represent comprehensive, patient-specific computational models that integrate multiple data sources, including medical imaging, to simulate individual physiology and pathology. These models increasingly leverage machine learning to personalize parameters based on patient-specific data and predict responses to potential interventions. Imaging provides essential anatomical and functional information for these models, while machine learning enables efficient parameter estimation and uncertainty quantification. Clinical applications include surgical planning, device design, treatment optimization, and outcome prediction, potentially supporting personalized interventions with reduced trial-and-error approaches. As computational capabilities advance, these digital representations will likely incorporate greater physiological detail and temporal dynamics, further enhancing their predictive value.

7.3. Advanced Model Architectures and Techniques

Transformer-based architectures, originally developed for natural language processing, have increasingly impacted medical image analysis by introducing self-attention mechanisms that capture long-range dependencies across image regions.

Vision Transformers (ViT) and their medical variants process images as sequences of patches, modeling relationships between distant image elements more effectively than convolutional approaches. These architectures have demonstrated particular value for whole-slide pathology images, volumetric scans, and multimodal integration where contextual information across the entire image is crucial. Hybrid architectures combining convolutional operations (for local feature extraction) with transformer blocks (for global context modeling) balance computational efficiency with representational capacity, addressing the scale and complexity of medical imaging data.

Generative models, including variational autoencoders (VAEs), generative adversarial networks (GANs), and diffusion models, learn to synthesize realistic medical images rather than merely analyzing existing ones. These approaches enable diverse applications including data augmentation (generating synthetic training examples), domain adaptation (translating images between acquisition protocols), anomaly detection (identifying deviations from learned normality), image enhancement (improving quality beyond traditional processing), and cross-modality synthesis (generating virtual representations of missing modalities). As generative capabilities advance, these models increasingly preserve clinically relevant features while manipulating image characteristics, supporting applications from privacy-preserving data sharing to radiation dose reduction through synthetic image generation.

Neuro-symbolic approaches combine deep learning with explicit symbolic reasoning, integrating the pattern recognition capabilities of neural networks with the interpretability and logical consistency of symbolic systems. In medical imaging, these hybrid approaches incorporate anatomical knowledge, clinical guidelines, and biological constraints into learning processes, potentially addressing limitations of pure data-driven methods. Applications include anatomy-aware segmentation (enforcing topological constraints), diagnostic reasoning (combining image features with clinical rules), and treatment planning (integrating image analysis with procedural knowledge). By formalizing domain knowledge alongside statistical learning, neuro-symbolic systems potentially achieve greater sample efficiency, interpretability, and alignment with clinical reasoning processes.

Graph neural networks (GNNs) provide a powerful framework for modeling relationships between image regions, capturing anatomical connectivity patterns and spatial dependencies beyond regular grid structures. In medical applications, GNNs represent anatomical structures as nodes connected by meaningful relationships, such as vascular networks, brain connectivity, cellular interactions, or organ adjacency. These approaches have demonstrated value in brain image analysis (modeling functional and structural connectivity), histopathology (capturing cellular relationships), and anatomical modeling (representing inter-organ dependencies). By explicitly modeling relational information rather than treating images as independent pixel arrays, GNNs potentially capture clinically relevant structural and functional relationships that may elude conventional architectures.

7.4. Clinical Translation and Implementation

Federated learning frameworks enable collaborative model training across institutions without sharing sensitive patient data, addressing privacy concerns while leveraging diverse datasets. Participating sites train models locally on their data, sharing only model updates rather than raw images. Aggregation servers combine these updates into consensus models that benefit from broad data diversity without centralizing protected health information. Advanced federated approaches incorporate differential privacy (adding noise to protect individual patient data), secure aggregation (cryptographic techniques preventing exposure of individual updates), and personalization mechanisms (adapting global models to local patient characteristics). These distributed learning paradigms potentially accelerate algorithm development while respecting privacy regulations and institutional data governance requirements.

Edge computing and on-device processing deploy machine learning algorithms directly on imaging equipment or local computing infrastructure rather than requiring cloud transmission. These approaches address latency constraints for time-sensitive applications, alleviate privacy concerns associated with data transfer, and enable deployment in settings with limited connectivity. Model optimization techniques including quantization, pruning, and architecture-specific optimizations enable sophisticated algorithms to run efficiently on resource-constrained devices. Clinical applications range from real-time guidance during interventional procedures to point-of-care diagnostic support in remote locations. As specialized hardware for machine learning inference becomes more widely available and energy-efficient, on-device processing will likely expand across the imaging workflow.

Human-AI collaboration frameworks recognize that machine learning systems complement rather than replace human expertise, designing interfaces and workflows that optimize this partnership. Effective collaboration models leverage respective strengths—algorithms excel at consistent pattern recognition across large datasets, while clinicians provide contextual understanding, handle exceptions, and integrate broader knowledge. Interface design considerations include appropriate confidence communication, attention direction to relevant findings, explanation generation tailored to expertise level, and smooth handoff mechanisms between algorithmic and human processing. Beyond technical performance, optimizing these collaborative aspects significantly influences clinical impact and adoption rates for machine learning systems in medical imaging.

Sustainability and scalability considerations increasingly influence implementation approaches as machine learning systems transition from research prototypes to clinical tools. Sustainable deployment addresses long-term maintenance requirements, adaptation to evolving clinical practices, monitoring for performance degradation, and efficient updating mechanisms. Scalable architecture accommodates growing data volumes, expanding use cases, and deployment across diverse healthcare settings with varying resource availability. Economic models for reimbursement, cost-effectiveness evaluation frameworks, and integration with existing healthcare financing systems further support widespread adoption and sustained utilization. As the field matures, these practical implementation aspects will likely receive greater attention alongside technical performance metrics.

7.5 Federated Learning in Medical Imaging

Federated Learning (FL) represents a paradigm shift in collaborative machine learning, enabling multiple decentralized institutions to collaboratively train AI models without directly exchanging sensitive patient data. This is particularly critical in healthcare, where stringent privacy regulations (e.g., HIPAA, GDPR) limit the feasibility of centralized data aggregation.

The Medical Open Network for AI (MONAI), an open-source initiative, has pioneered the application of federated learning in medical imaging. In a notable case study, MONAI facilitated a collaborative tumor segmentation project involving 10 geographically distributed hospitals. Each hospital retained its local patient imaging data (MRI, CT scans) within its secured infrastructure while contributing to the collective model training process through the MONAI Federated Learning framework. Key Features of the MONAI Federated Learning Approach:

- Privacy-Preserving Architecture: Model parameters, rather than raw patient data, are shared and aggregated using secure protocols.
- Adaptive Learning: Local models are fine-tuned on site-specific data, enhancing model generalizability across diverse patient populations.
- Performance: The federated model achieved comparable or superior segmentation accuracy relative to centralized models, while maintaining data privacy compliance.

Clinical Impact:

- Accelerated development of robust tumor segmentation models.
- Preservation of patient confidentiality across institutions.

- Establishment of a scalable framework for future multi-institutional collaborations.

7.6 Digital Twins in Personalized Medicine

Digital Twins in healthcare refer to virtual replicas of individual patients that simulate their physiological states, disease progression, and responses to treatments. This emerging concept leverages real-time data integration from wearable devices, imaging studies, laboratory tests, and electronic health records (EHRs).

By constructing a digital twin of a patient's cardiovascular system, clinicians can simulate various treatment scenarios, evaluate risks, and optimize therapeutic interventions personalized to the patient's unique profile.

Benefits of Digital Twins:

- Personalized treatment planning.
- Predictive modeling of disease evolution.
- Real-time monitoring and intervention optimization.

Future Potential:

- Integration with AI-driven diagnostic systems.
- Dynamic updates based on continuous patient data streams.
- Enabling precision medicine at the individual patient level.

8. CONCLUSION

Machine learning has fundamentally transformed medical image analysis, enabling unprecedented capabilities in disease detection, precise anatomical delineation, quantitative assessment, and treatment planning across imaging modalities. This chapter explores theoretical foundations, methodological approaches, clinical applications, and emerging trends in this rapidly evolving field, highlighting both remarkable achievements and persistent challenges. The integration of advanced deep learning architectures with domain-specific adaptations has yielded systems that approach or occasionally surpass human performance in specific tasks, from pulmonary nodule detection to cardiac function assessment. These technical accomplishments have been translated into clinical implementations that augment diagnostic workflows, enhance treatment precision, and improve efficiency across healthcare settings. Applications span the entire imaging chain from acquisition optimization through enhanced reconstruction, automated analysis, and decision support, with growing impact across clinical specialties. Despite this progress, significant challenges remain in developing robust, generalizable, and trustworthy systems suitable for diverse clinical environments. Data limitations, including scarcity of expert annotations and representation biases, constrain model development and evaluation. Technical hurdles in interpretability, uncertainty quantification, and domain adaptation complicate clinical translation and regulatory approval. Implementation challenges spanning workflow integration, performance monitoring, and sustainability considerations affect real-world impact beyond laboratory settings. Ethical considerations regarding algorithmic fairness, appropriate autonomy levels, and equitable access further complicate responsible deployment. Emerging approaches offer promising directions for addressing these challenges. Self-supervised and federated learning paradigms leverage unlabeled data and distributed computation to overcome annotation scarcity while preserving privacy. Multimodal integration and neuro-symbolic architectures incorporate diverse information sources and domain knowledge, potentially improving robustness and clinical relevance. Human-centered design approaches optimize collaborative workflows between algorithms and clinicians, leveraging complementary strengths rather than pursuing full automation. The future of machine learning in medical imaging will likely feature increasingly personalized approaches that adapt to individual patient characteristics, institution-specific workflows, and clinician preferences. Systems will evolve from static algorithms to learning companions that continuously improve through clinical feedback while maintaining reliability and transparency. Integration across traditionally separated disciplines—radiology, pathology, genomics, clinical informatics—will accelerate, supported by machine learning approaches that bridge modality-specific limitations and synthesize comprehensive patient representations. Realizing this potential requires continued interdisciplinary collaboration among technical experts, clinical practitioners, regulatory specialists, and patient advocates. Coordinated efforts to develop representative datasets, establish rigorous evaluation methodologies, design human-centered implementations, and create appropriate governance frameworks will determine whether machine learning truly transforms healthcare delivery or remains limited to isolated applications. With thoughtful development prioritizing clinical needs over technical novelty, these powerful methods can enhance diagnostic accuracy, treatment personalization, operational efficiency, and ultimately patient outcomes across healthcare systems worldwide.

REFERENCES

Abdou, M. A. (2022). Literature review: Efficient deep neural networks techniques for medical image analysis. *Neural Computing & Applications*, *34*(8), 5791–5812. DOI: 10.1007/s00521-022-06960-9

Ali, S., Li, J., Pei, Y., Khurram, R., Rehman, K. U., & Rasool, A. B. (2021). State-of-the-art challenges and perspectives in multi-organ cancer diagnosis via deep learning-based methods. *Cancers (Basel)*, *13*(21), 5546. DOI: 10.3390/cancers13215546 PMID: 34771708

Allgaier, J., & Pryss, R. (2024). Cross-validation visualized: A narrative guide to advanced methods. *Machine Learning and Knowledge Extraction*, *6*(2), 1378–1388. DOI: 10.3390/make6020065

Alloghani, M., Al-Jumeily, D., Mustafina, J., Hussain, A., & Aljaaf, A. J. (2020). A systematic review on supervised and unsupervised machine learning algorithms for data science. *Supervised and unsupervised learning for data science*, 3-21.

Alqahtani, F. F. (2023). SPECT/CT and PET/CT, related radiopharmaceuticals, and areas of application and comparison. *Saudi Pharmaceutical Journal : SPJ : the Official Publication of the Saudi Pharmaceutical Society*, *31*(2), 312–328. DOI: 10.1016/j.jsps.2022.12.013 PMID: 36942267

Anwar, S. M., Majid, M., Qayyum, A., Awais, M., Alnowami, M., & Khan, M. K. (2018). Medical image analysis using convolutional neural networks: A review. *Journal of Medical Systems*, *42*(11), 1–13. DOI: 10.1007/s10916-018-1088-1 PMID: 30298337

Atmakuru, A., Chakraborty, S., Faust, O., Salvi, M., Barua, P. D., Molinari, F., & Homaira, N. (2024). Deep learning in radiology for lung cancer diagnostics: A systematic review of classification, segmentation, and predictive modeling techniques. *Expert Systems with Applications*, *255*, 124665. DOI: 10.1016/j.eswa.2024.124665

Bushong, S. C., & Clarke, G. (2003). *Magnetic resonance imaging: physical and biological principles*. Elsevier Health Sciences.

Chen, H. L., Yang, B., Wang, G., Wang, S. J., Liu, J., & Liu, D. Y. (2012). Support vector machine based diagnostic system for breast cancer using swarm intelligence. *Journal of Medical Systems*, *36*(4), 2505–2519. DOI: 10.1007/s10916-011-9723-0 PMID: 21537848

Cierniak, R. (2011). *X-ray computed tomography in biomedical engineering*. Springer Science & Business Media. DOI: 10.1007/978-0-85729-027-4

Cunningham, P., & Delany, S. J. (2021). K-nearest neighbour classifiers-a tutorial. *ACM Computing Surveys*, *54*(6), 1–25. DOI: 10.1145/3459665

Dougherty, G. (2009). *Digital image processing for medical applications*. Cambridge University Press. DOI: 10.1017/CBO9780511609657

El-Latif, A. A. A., Chelloug, S. A., Alabdulhafith, M., & Hammad, M.AbdEl-Latif. (2023). Accurate detection of Alzheimer's disease using lightweight deep learning model on MRI data. *Diagnostics (Basel)*, *13*(7), 1216. DOI: 10.3390/diagnostics13071216 PMID: 37046434

Elangovan, A., & Jeyaseelan, T. (2016, February). Medical imaging modalities: a survey. In *2016 International Conference on emerging trends in engineering, technology and science (ICETETS)* (pp. 1-4). ieee.

Fournier, L., de Geus-Oei, L. F., Regge, D., Oprea-Lager, D. E., D'Anastasi, M., Bidaut, L., Bäuerle, T., Lopci, E., Cappello, G., Lecouvet, F., Mayerhoefer, M., Kunz, W. G., Verhoeff, J. J. C., Caruso, D., Smits, M., Hoffmann, R.-T., Gourtsoyianni, S., Beets-Tan, R., Neri, E., & Caramella, C. (2022). Twenty years on: RECIST as a biomarker of response in solid tumours an EORTC imaging group–ESOI joint paper. *Frontiers in Oncology*, *11*, 800547. DOI: 10.3389/fonc.2021.800547 PMID: 35083155

Fu, Y., Lei, Y., Wang, T., Curran, W. J., Liu, T., & Yang, X. (2021). A review of deep learning based methods for medical image multi-organ segmentation. *Physica Medica*, *85*, 107–122. DOI: 10.1016/j.ejmp.2021.05.003 PMID: 33992856

Gonem, S., Janssens, W., Das, N., & Topalovic, M. (2020). Applications of artificial intelligence and machine learning in respiratory medicine. *Thorax*, *75*(8), 695–701. DOI: 10.1136/thoraxjnl-2020-214556 PMID: 32409611

Hammad, M., Bakrey, M., Bakhiet, A., Tadeusiewicz, R., Abd El-Latif, A. A., & Pławiak, P. (2022). A novel end-to-end deep learning approach for cancer detection based on microscopic medical images. *Biocybernetics and Biomedical Engineering*, *42*(3), 737–748. DOI: 10.1016/j.bbe.2022.05.009

Hammad, M., ElAffendi, M., Asim, M., Abd El-Latif, A. A., & Hashiesh, R. (2024). Automated lung cancer detection using novel genetic TPOT feature optimization with deep learning techniques. *Results in Engineering*, *24*, 103448. DOI: 10.1016/j.rineng.2024.103448

Hammad, M., ElAffendi, M., Ateya, A. A., & Abd El-Latif, A. A. (2023). Efficient brain tumor detection with lightweight end-to-end deep learning model. *Cancers (Basel)*, *15*(10), 2837. DOI: 10.3390/cancers15102837 PMID: 37345173

Hammad, M., Meshoul, S., Dziwiński, P., Pławiak, P., & Elgendy, I. A. (2022). Efficient lightweight multimodel deep fusion based on ECG for arrhythmia classification. *Sensors (Basel)*, *22*(23), 9347. DOI: 10.3390/s22239347 PMID: 36502049

Hammad, M., Pławiak, P., ElAffendi, M., El-Latif, A. A. A., & Latif, A. A. A. (2023). Enhanced deep learning approach for accurate eczema and psoriasis skin detection. *Sensors (Basel)*, *23*(16), 7295. DOI: 10.3390/s23167295 PMID: 37631831

Hammad, M., Tawalbeh, L. A., Iliyasu, A. M., Sedik, A., Abd El-Samie, F. E., Alkinani, M. H., & Abd El-Latif, A. A. (2022). Efficient multimodal deep-learning-based COVID-19 diagnostic system for noisy and corrupted images. *Journal of King Saud University. Science*, *34*(3), 101898. DOI: 10.1016/j.jksus.2022.101898 PMID: 35185304

Hiriyannaiah, H. P. (1997). X-ray computed tomography for medical imaging. *IEEE Signal Processing Magazine*, *14*(2), 42–59. DOI: 10.1109/79.581370

Hussain, D., Al-Masni, M. A., Aslam, M., Sadeghi-Niaraki, A., Hussain, J., Gu, Y. H., & Naqvi, R. A. (2024). Revolutionizing tumor detection and classification in multimodality imaging based on deep learning approaches: Methods, applications and limitations. *Journal of X-Ray Science and Technology*, *32*(4), 857–911. DOI: 10.3233/XST-230429 PMID: 38701131

Islam, S. M. S., Nasim, M. A. A., Hossain, I., Ullah, D. M. A., Gupta, D. K. D., & Bhuiyan, M. M. H. (2023). Introduction of medical imaging modalities. In *Data driven approaches on medical imaging* (pp. 1–25). Springer Nature Switzerland. DOI: 10.1007/978-3-031-47772-0_1

Jiang, H., Diao, Z., Shi, T., Zhou, Y., Wang, F., Hu, W., Zhu, X., Luo, S., Tong, G., & Yao, Y. D. (2023). A review of deep learning-based multiple-lesion recognition from medical images: Classification, detection and segmentation. *Computers in Biology and Medicine*, *157*, 106726. DOI: 10.1016/j.compbiomed.2023.106726 PMID: 36924732

Johnston, B., Jones, A., & Kruger, C. (2019). *Applied Unsupervised Learning with Python: Discover hidden patterns and relationships in unstructured data with Python*. Packt Publishing Ltd.

Jordan, M. I., & Mitchell, T. M. (2015). Machine learning: Trends, perspectives, and prospects. *Science*, *349*(6245), 255–260. DOI: 10.1126/science.aaa8415 PMID: 26185243

Ker, J., Wang, L., Rao, J., & Lim, T. (2017). Deep learning applications in medical image analysis. *IEEE Access : Practical Innovations, Open Solutions*, *6*, 9375–9389. DOI: 10.1109/ACCESS.2017.2788044

Kumar, B., Dikshit, O., Gupta, A., & Singh, M. K. (2020). Feature extraction for hyperspectral image classification: A review. *International Journal of Remote Sensing*, *41*(16), 6248–6287. DOI: 10.1080/01431161.2020.1736732

Kumar, S., Kumar, H., Kumar, G., Singh, S. P., Bijalwan, A., & Diwakar, M. (2024). A methodical exploration of imaging modalities from dataset to detection through machine learning paradigms in prominent lung disease diagnosis: A review. *BMC Medical Imaging*, *24*(1), 30. DOI: 10.1186/s12880-024-01192-w PMID: 38302883

Li, M., Jiang, Y., Zhang, Y., & Zhu, H. (2023). Medical image analysis using deep learning algorithms. *Frontiers in Public Health*, *11*, 1273253. DOI: 10.3389/fpubh.2023.1273253 PMID: 38026291

. Li, X., Li, M., Yan, P., Li, G., Jiang, Y., Luo, H., & Yin, S. (2023). Deep learning attention mechanism in medical image analysis: Basics and beyonds. *International Journal of Network Dynamics and Intelligence*, 93-116.

Li, X., Zhang, L., Yang, J., & Teng, F. (2024). Role of artificial intelligence in medical image analysis: A review of current trends and future directions. *Journal of Medical and Biological Engineering*, *44*(2), 231–243. DOI: 10.1007/s40846-024-00863-x

Litjens, G., Ciompi, F., Wolterink, J. M., de Vos, B. D., Leiner, T., Teuwen, J., & Išgum, I. (2019). State-of-the-art deep learning in cardiovascular image analysis. *JACC: Cardiovascular Imaging*, *12*(8 Part 1), 1549–1565. DOI: 10.1016/j.jcmg.2019.06.009 PMID: 31395244

Maćkiewicz, A., & Ratajczak, W. (1993). Principal components analysis (PCA). *Computers & Geosciences*, *19*(3), 303–342. DOI: 10.1016/0098-3004(93)90090-R

Maglogiannis, I. G., & Zafiropoulos, E. P. (2004). Characterization of digital medical images utilizing support vector machines. *BMC Medical Informatics and Decision Making*, *4*(1), 1–9. DOI: 10.1186/1472-6947-4-4 PMID: 15113418

Modi, S. K., Singla, S., Modi, P., & Kaur, G. K. (2023, October). Advancements in Neurodegenerative Disease Diagnosis and Prediction: A Machine Learning Approach. In *International Health Informatics Conference* (pp. 257-272). Singapore: Springer Nature Singapore.

. Mucherino, A., Papajorgji, P. J., Pardalos, P. M., Mucherino, A., Papajorgji, P. J., & Pardalos, P. M. (2009). K-nearest neighbor classification. *Data mining in agriculture*, 83-106.

Naidu, G., Zuva, T., & Sibanda, E. M. (2023, April). A review of evaluation metrics in machine learning algorithms. In *Computer science on-line conference* (pp. 15–25). Springer International Publishing. DOI: 10.1007/978-3-031-35314-7_2

Najjar, R. (2023). Redefining radiology: A review of artificial intelligence integration in medical imaging. *Diagnostics (Basel), 13*(17), 2760. DOI: 10.3390/diagnostics13172760 PMID: 37685300

Nananukul, N., Soltanian-Zadeh, H., & Rostami, M. (2024). Multi-source data integration for segmentation of unannotated mri images. *IEEE Journal of Biomedical and Health Informatics*, 1–8. DOI: 10.1109/JBHI.2024.3422250 PMID: 38954567

Ou, X., Chen, X., Xu, X., Xie, L., Chen, X., Hong, Z., & Yang, H. (2021). *Recent development in x-ray imaging technology: Future and challenges*. Research.

Parmar, C., Barry, J. D., Hosny, A., Quackenbush, J., & Aerts, H. J. (2018). Data analysis strategies in medical imaging. *Clinical Cancer Research, 24*(15), 3492–3499. DOI: 10.1158/1078-0432.CCR-18-0385 PMID: 29581134

Pinto-Coelho, L. (2023). How artificial intelligence is shaping medical imaging technology: A survey of innovations and applications. *Bioengineering (Basel, Switzerland), 10*(12), 1435. DOI: 10.3390/bioengineering10121435 PMID: 38136026

Pouyanfar, S., Sadiq, S., Yan, Y., Tian, H., Tao, Y., Reyes, M. P., Shyu, M.-L., Chen, S.-C., & Iyengar, S. S. (2018). A survey on deep learning: Algorithms, techniques, and applications. *ACM Computing Surveys, 51*(5), 1–36. DOI: 10.1145/3234150

Qi, G. J., & Luo, J. (2020). Small data challenges in big data era: A survey of recent progress on unsupervised and semi-supervised methods. *IEEE Transactions on Pattern Analysis and Machine Intelligence, 44*(4), 2168–2187. DOI: 10.1109/TPAMI.2020.3031898 PMID: 33074801

Rahman, H., Aoun, N. B., Bukht, T. F. N., Ahmad, S., Tadeusiewicz, R., Pławiak, P., & Hammad, M. (2025). Automatic Liver Tumor Segmentation of CT and MRI Volumes Using Ensemble ResUNet-InceptionV4 Model. *Information Sciences, 704*, 121966. DOI: 10.1016/j.ins.2025.121966

Raza, K., & Singh, N. K. (2021). A tour of unsupervised deep learning for medical image analysis. *Current Medical Imaging, 17*(9), 1059–1077. DOI: 10.2174/1573405617666210127154257 PMID: 33504314

Rogowska, J. (2009). Overview and fundamentals of medical image segmentation. *Handbook of medical image processing and analysis*, 73-90.

Rueckert, D., & Schnabel, J. A. (2010). Medical image registration. In *Biomedical image processing* (pp. 131–154). Springer Berlin Heidelberg. DOI: 10.1007/978-3-642-15816-2_5

Sakr, A. S., Pławiak, P., Tadeusiewicz, R., Pławiak, J., Sakr, M., & Hammad, M. (2023). ECG-COVID: An end-to-end deep model based on electrocardiogram for COVID-19 detection. *Information Sciences*, *619*, 324–339. DOI: 10.1016/j.ins.2022.11.069 PMID: 36415325

Salau, A. O., & Jain, S. (2019, March). Feature extraction: a survey of the types, techniques, applications. In *2019 international conference on signal processing and communication (ICSC)* (pp. 158-164). IEEE. DOI: 10.1109/ICSC45622.2019.8938371

Sen, P. C., Hajra, M., & Ghosh, M. (2020). Supervised classification algorithms in machine learning: A survey and review. In *Emerging Technology in Modelling and Graphics:Proceedings of IEM Graph 2018* (pp. 99-111). Springer Singapore. DOI: 10.1007/978-981-13-7403-6_11

Shamshad, F., Khan, S., Zamir, S. W., Khan, M. H., Hayat, M., Khan, F. S., & Fu, H. (2023). Transformers in medical imaging: A survey. *Medical Image Analysis*, *88*, 102802. DOI: 10.1016/j.media.2023.102802 PMID: 37315483

Shehab, M., Abualigah, L., Shambour, Q., Abu-Hashem, M. A., Shambour, M. K. Y., Alsalibi, A. I., & Gandomi, A. H. (2022). Machine learning in medical applications: A review of state-of-the-art methods. *Computers in Biology and Medicine*, *145*, 105458. DOI: 10.1016/j.compbiomed.2022.105458 PMID: 35364311

Sherstinsky, A. (2020). Fundamentals of recurrent neural network (RNN) and long short-term memory (LSTM) network. *Physica D. Nonlinear Phenomena*, *404*, 132306. DOI: 10.1016/j.physd.2019.132306

Showrov, A. A., Aziz, M. T., Nabil, H. R., Jim, J. R., Kabir, M. M., Mridha, M. F., & Shin, J. (2024). Generative adversarial networks (GANs) in medical imaging: Advancements, applications and challenges. *IEEE Access : Practical Innovations, Open Solutions*.

Suganyadevi, S., Seethalakshmi, V., & Balasamy, K. (2022). A review on deep learning in medical image analysis. *International Journal of Multimedia Information Retrieval*, *11*(1), 19–38. DOI: 10.1007/s13735-021-00218-1 PMID: 34513553

Szabo, T. L. (2013). *Diagnostic ultrasound imaging: inside out*. Academic press.

Thakur, G. K., Thakur, A., Kulkarni, S., Khan, N., & Khan, S. (2024). Deep learning approaches for medical image analysis and diagnosis. *Cureus*, *16*(5). Advance online publication. DOI: 10.7759/cureus.59507 PMID: 38826977

Tsuneki, M. (2022). Deep learning models in medical image analysis. *Journal of Oral Biosciences*, *64*(3), 312–320. DOI: 10.1016/j.job.2022.03.003 PMID: 35306172

Withey, D. J., & Koles, Z. J. (2008). A review of medical image segmentation: Methods and available software. *International Journal of Bioelectromagnetism*, *10*(3), 125–148.

Yadav, A. K., Kumar, N., & Tripathi, A. Machine Learning Approaches in Neurodegenerative Diseases. In *Synaptic Plasticity in Neurodegenerative Disorders* (pp. 250–264). CRC Press.

Yeasmin, M. N., Al Amin, M., Joti, T. J., Aung, Z., & Azim, M. A. (2024). Advances of AI in image-based computer-aided diagnosis: A review. *Array (New York, N.Y.)*, *23*, 100357. DOI: 10.1016/j.array.2024.100357

Zheng, A., & Casari, A. (2018). *Feature engineering for machine learning: principles and techniques for data scientists*. O'Reilly Media, Inc.

Zhou, S. K., Greenspan, H., Davatzikos, C., Duncan, J. S., Van Ginneken, B., Madabhushi, A., Prince, J. L., Rueckert, D., & Summers, R. M. (2021). A review of deep learning in medical imaging: Imaging traits, technology trends, case studies with progress highlights, and future promises. *Proceedings of the IEEE*, *109*(5), 820–838. DOI: 10.1109/JPROC.2021.3054390 PMID: 37786449

Chapter 6
Optimising Healthcare Outcomes With Prompt Engineering: A Transformative Approach

Hemlata Parmar
https://orcid.org/0009-0009-1438-5427
Manipal University Jaipur, India

Utsav Krishan Murari
https://orcid.org/0009-0007-1606-6775
Sharda University, India

ABSTRACT

This chapter focuses on identifying the potential of prompt engineering to revolutionise health care with the help of applied research and case studies. With the help of new generative models, such as GPT-4, healthcare specialists work on clinical problems with the help of better diagnostics, improved interaction with patients, and more effective treatment plans. This chapter describes the strategies for calibrating prompts to produce context-aware, reliable responses that can support real-time decision-making in clinical practice. Prompt engineering enhances the health of clinicians since the software manages routine jobs. This chapter also looks at its usage in medical education using AI-based case scenarios, which help grasp intricate clinical ideas and reasoning abilities. Thus, prompt engineering can help healthcare professionals increase diagnostic performance, engage patients more effectively, and optimise the work process. This chapter establishes the need for prompt engineering as one of the most essential tools in the current and future healthcare field.

DOI: 10.4018/979-8-3373-0548-6.ch006

1. INTRODUCTION

It is important to mention that artificial intelligence (AI) is key in healthcare, providing operations from predictive analytics to personalised medicine. Biomedical Applications using Large Language Models (LLMs) like GPT-4 process such massive biomedical data and generate actionable insights. Thus, examples like the use of LLMs in aiding diagnostic precision in emergency medicine show that LLMs carry high stakes in healthcare, where the accuracy of responses matters, and the jump to scalable inference has yet to materialise. Prompts designed as intended also allow the practitioner to derive clinically appropriate outputs from the AI systems, closing the gap between raw data and actionable intelligence. A recent and highly effective development is the field of prompt engineering, a fundamental approach for increasing the performance of large language models (LLMs). The process of constructing and organising inputs to AI systems to generate contextually relevant, specific-purpose responses is called prompt engineering. It cannot be over-emphasised in healthcare, where the consequences are dire, and the factors are numerous and interdependent. Given that GPT4 and other LLMs are available to clinicians and researchers, they can fine-tune these models to enhance decision-making or diagnostic capabilities or provide better patient information. As a result, the healthcare ecosystem is evolving, making it more efficient, data-driven, and focused on patients. Artificial intelligence (AI) plays a critical role in healthcare in areas as diverse as predictive analytics and personalised medicine to hospital operational efficiencies. Under this paradigm, LLMs act as a game changer that can process enormous amounts of unstructured biomedical data and generate actionable insights by synthesising them. At the same time, these systems are intrinsically related to the quality and specificity of the input they are provided. Practitioners can use the full potential of LLMs by crafting good prompts; here, the act of 'good prompts' enables the act of information retrieval, information interpretation, and information application in real-world clinical scenarios (Brown et al., 2020).

1.1 The Role of LLMs in Healthcare

Thanks to LLMs, clinical decision-making has been supported, dense literature has been summarised, and patient communication has been facilitated in healthcare. For example, LLMs delivered rapid, accurate diagnoses to enable rapid, accurate treatment during emergencies, as in the case of emergency medicine. So, structured LLM inputs become inputs with patient history and contextual cues, which means that healthcare professionals get targeted and actionable outputs. Large language models, GPT-4 being at the forefront of AI capabilities. These pre-trained models on diverse datasets can generate human-like text responses, translate languages or simulate reasoning processes. LLMs have been game changers in Healthcare by supporting clinical decision-making, summarising dense medical literature and aiding in patient communication. However, this generality of training becomes a problem when addressing specialised medical tasks. In clinical settings, the outputs of LLMs with proper contextual guidance may be more specific and reliable. However, this is where prompt engineering comes into play, allowing users to customise interactions to guarantee that AI systems provide accurate, context-sensitive responses. Take, for example, a prompt to make a differential diagnosis on a patient presenting with fever and rash – the prompt must contain details about the patient's medical history, recent travel history, co-morbidities, etc. Instead, carefully structuring a prompt guarantees that the prompted AI will consider clinically pertinent parameters and provide a specific and actionable output instead of an incomplete or vast output for a generic prompt.

1.2 Foundations of Prompt Engineering

By allowing natural language interaction with AI, prompt engineering enables non-technical users such as clinicians to utilise AI. Unlike traditional programming, which requires detailed coding, prompt engineering makes data querying intuitive and maximises AI output. This democratisation of access to AI enables researchers and clinicians to apply. AI tools to real-world problems without needing technical expertise. At its core, prompt engineering is a methodology for the structured design of inputs to maximise the utility of outputs from AI. Therefore, in healthcare, the flow of medical terminology, contextual cues, and domain-specific knowledge are combined in prompts. Unlike traditional programming paradigms, which require explicit coding, prompt engineering allows natural language to be the primary interface for LLMs, thereby allowing non-technical users such as clinicians and researchers to use LLM effectively.

Figure 1. Prompt Engineering in Healthcare (Brown, T. et al. (2020))

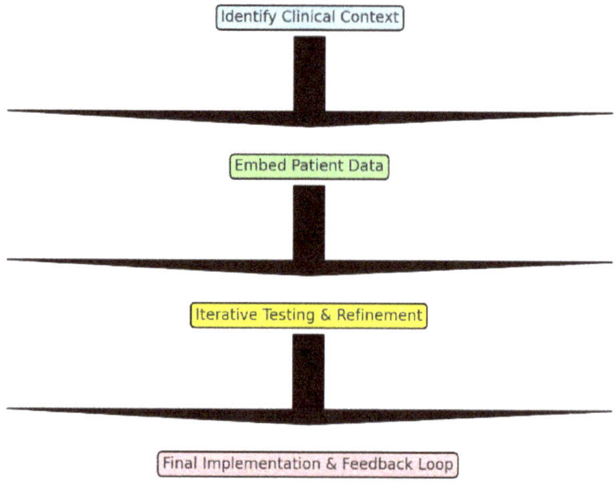

1.3 Techniques for Effective Prompt Design

Several techniques underpin effective, prompt engineering in healthcare:

- **Iterative Refinement:** Iterative Refinement includes trying and adapting prompts to increase clarity and relevance. Iterative refinement ensures that the AI system will really and truly understand your intent behind a query rather than the wording itself, making for far more accurate and actionable outputs.
- **Few-Shot and Zero-Shot Learning**: Examples of the desired output are embedded within the prompt to prompt the AI where to go. This is quite interesting for tasks like summarising patient records or generating treatment plans. On the other hand, zero-shot learning involves the AI interpreting and responding to novel prompts using its pre-trained capabilities, but this requires careful input composition to avoid ambiguity.

Table 1. Comparison of Few-Shot and Zero-Shot learning techniques based on various criteria

S.No.	Criteria	Few-Shot Learning	Zero-Shot Learning
1	Accuracy	Higher accuracy with examples	Lower initial accuracy
2	Data Requirements	Requires example inputs	No examples required
3	Applicability	Useful for complex tasks	Useful for novel tasks
4	Example	Summarising patient records	Interpreting unknown symptoms

- **Context Embedding**: To guarantee that AI provides clinically relevant and case-specific outputs, users can embed the appropriate contextual information into prompts—such as patient demographics, symptoms, and laboratory results.

1.4 Challenges in Prompt Engineering

Designing prompts that lead to accurate and unbiased responses is a problem of prompt engineering in healthcare. For instance, a fuzzy query such as the best cough treatment. May yield incomplete results. Another way to envision it becomes 'What are treatment options for a continuous cough in a 45-year-old smoker with chronic obstructive pulmonary disease?' This is a necessity for precision. It is not AI substituting for the judgment of a physician, and it augments the physician's judgment. Although the scope of prompt engineering is enormous, it comes with some difficulties. Designing prompts that elicit accurate, unbiased responses from LLMs is hardly sufficient to swap out physicians with LLMs. An ambiguous or alarming prompt structure can result in misleading outputs inappropriate for patient safety. In addition, the dependence on pre-trained models poses a problem of generalising AI systems to different patient cohorts and clinical situations such that refinement and validation of the prompt design techniques are ongoing (Mehrabi et al., 2021).

Figure 2. Challenges and Ethical Considerations (Mehrabi. et al. (2021))

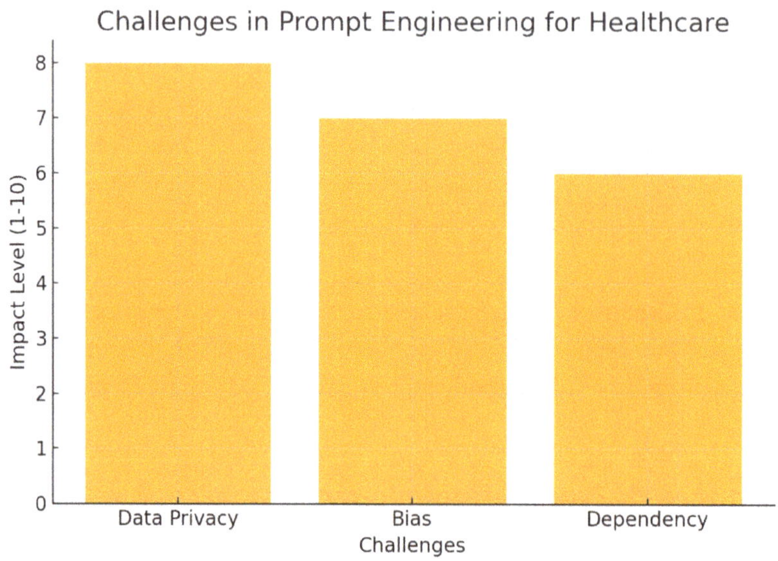

1.5 Applications of Prompt Engineering in Healthcare

The versatility of prompt engineering is evident in its diverse applications across healthcare domains:

Figure 3. Key Features of Large Language Models (LLMs) in Healthcare (Smith, H. et al. (2023))

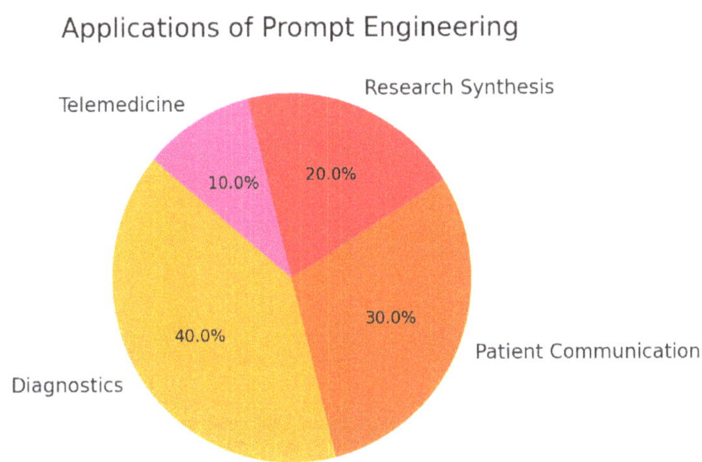

- **Clinical Decision Support Systems (CDSS)**: Using prompt engineering, LLMs can be intelligent assistants in clinical decision-making. With the inputs structured to contain patient-specific data, clinicians can use AI to generate differential diagnoses, suggest evidence-based treatments, and indicate possible complications. For example, 'What are the most likely causes of persistent cough in a 45-year-old smoker' with a history of chronic obstructive pulmonary disease? It can create a list of prioritised probability conditions, which will help clinicians cut short diagnostic possibilities (Johnson et al., 2023).
- **Biomedical Research and Literature Synthesis:** For research, LLMs are instrumental because they can search through much medical literature. The researchers can extract information from the unstructured data, detect trends, and make assumptions through quick engineering.
- **Patient Education and Communication:** Another fundamental of reasonable care is patient education on the disease and available treatment. We can develop educational materials for literacy levels, languages, and cultures through prompt engineering. For instance, the prompts can provide simple descriptions of complex medical procedures or build individualised consumption of medication behaviours.

- **Telemedicine and Remote Monitoring**: Telemedicine platforms are necessary to serve remote and underserved populations in digital healthcare. This offers significant improvements to the functionality of these platforms by having AI triage patients, summarise virtual consultations, and make follow-up recommendations. For instance, a prompt like generating a follow-up care plan for a patient with uncontrolled hypertension will do away with the telemedicine workflow.

Table 2. Key Features of Large Language Models (LLMs) in Healthcare

S.No.	Feature	Description	Examples in Healthcare
1	Natural Language Understanding	Ability to process and interpret human language	Summarising medical records
2	Knowledge Synthesis	Aggregating and analysing large datasets	Generating research summaries
3	Language Translation	Translating medical content across languages	Improving communication in multilingual settings
4	Clinical Reasoning	Simulating diagnostic and treatment decision-making	Differential diagnosis tools

- **Case Study 1: Emergency Decision Support**

To assist with differential diagnosis, an AI-powered system was combined with enhanced prompt engineering and integrated into a hospital's emergency department. It made triage 30% more efficient by infusing structured prompts in the form of patient demographics, medical history and symptoms. In time-sensitive environments where misdiagnosis could prove fatal, the specific prompts that prompted clinicians to review the patient's record helped avoid such misdiagnoses, as reported by clinicians (Smith, Brown, & Rivera, 2023).

- **Case Study 2: Oncology Treatment Planning**

LLMs were also used in a cancer care facility to design personalised treatment plans. Prompt-engineered queries were used to analyse patient genetic profiles and historical medical profiles and make recommendations for therapeutic approaches tailored to patients. The integration cut through 40 per cent of the time required for treatment planning and aligned with the most recent oncological guidelines (Lee et al., 2021).

1.6 Ethical Considerations in Prompt Engineering

Deploying prompt engineering in healthcare is not an ethical minefield. One of the biggest concerns is the bias in the AI outputs. LLMs can be biased from their training datasets or the prompts design. For instance, if a prompt causes a treatment recommendation to be discriminatory or unavailable to some populations (Mehrabi et al., 2021), it inadvertently excludes particular socioeconomic factors when they should have been considered. Rigorous validation processes and diverse datasets should be used when addressing these biases. Another critical concern is data privacy and security. When applied to healthcare, applications of AI will deal with sensitive patient information that necessitates strict compliance with the Health Insurance Portability and Accountability Act (HIPAA). To prevent breaches, prompt engineering must ensure that patient data embedded in inputs is anonymised.

1.7 The Future of Prompt Engineering in Healthcare

Since AI technology is getting improved, the role of prompt engineering in healthcare will increase. Future developments may include:

- **Multi-Modal Integration**: LLMs could gain access to a more robust context and thus respond more appropriately if the data from other modalities, like medical images, genomic sequences, or electronic health records, can be integrated into the prompt.
- **Personalised Medicine**: With prompt engineering, implementing personalised medicine could be supported by AI, generating more tailored treatment recommendations using a patient's genetic profile, lifestyle and clinical history.
- **Global Health Applications**: Spunky prompts could be simplified, and prestructured healthcare professionals in low-resource settings could use AI for diagnostics and treatment planning, minimising global health inequalities.
- **Interoperability with Emerging Technologies**: Integrating LLMs into robotics and the Internet of Things (IoT) devices could enable fully automated patient monitoring, surgical assistance, and realtime decision-making, thus expanding prompt engineering capabilities.

Prompt engineering is reshaping the confluence of AI and healthcare by allowing practitioners to leverage LLMs to do nuanced, context-aware, and consequential things. This methodology can transform clinical decision-making and improve research capabilities and patient education, revolutionising healthcare delivery. It holds full promise and requires addressing bias, privacy, and usability challenges. The evolution will shape the future of Artificial intelligence in healthcare as it relates to prompt engineering with emerging technologies and commitment to ethical standards.

2. CORE METHODOLOGIES OF PROMPT ENGINEERING IN HEALTHCARE

Artificial intelligence (AI) has recently been integrated into healthcare, bringing about massive changes in clinical, administrative, and patient-facing domains. However, the performance of an AI system is highly dependent on how well it is prompted towards generating accurate and contextually appropriate outputs, a task commonly known as prompt engineering. Crafting these inputs so that they can direct the AI models to respond as they would, within the notions of clinical standards, evidence-based practices, and the demands of a healthcare context, is the methodology of this thesis. The strategies for prompt engineering are OK tuning prompts for clarity and relevance, iterative testing for refining performance, and embedding domain knowledge for precision.

2.1 Fine-Tuning Prompts: Optimizing Clarity and Relevance

The actual work of practical prompt engineering is mainly in tuning the prompts. This methodology, therefore, guarantees the network that the input data into the AI system is correct and consistent so that the output is credible and coherent. The practice of altering the prompt's length and/or the degree to which it is detailed is called prompt engineering. This way, inputs provided to the AI are standard, not biased, and systematic in order to result in credible outcomes.

- **Adjusting Prompt Length and Specificity:** The length of the input prompt and the level of detail in the input prompt affect the result of the AI system's output. The inputs are pretty imprecise and short, which makes it difficult to understand what kind of information the model has to provide, so the model provides very general information. On the other hand, long prompts provide a direction but may provide it with noise or some unwanted data (Brown et al., 2020). This is especially the case because the decisions made within clinical practice in healthcare careers are critical, and these prompts must be brief and pertinent. For instance, what may be the causes of chest pain? This may lead to what might be termed over-diagnosis but a more directed question such as 'What could be the cause of chest pain in a 55-year-old male with a history of hypertension and smoking?' It brings a general result but a more specific question, such as what may be the cause of chest pain in a 55-year-old male with hypertension and a smoker. It offers information that enhances the clinical implications of the result while narrowing the field of view. Dynamic prompt optimisation has recently been indicated as crucial. We also establish that while studies reveal that prompts designed for task complexity are far more effective, the relationship between usability and shape is not only complex but also highly nonlinear, as, for instance, the accuracy of diagnostic applications is boosted when particular attention is paid to such factors as patient's demographics, the duration of the symptoms, and other clinical characteristics (Jiang et al., 2022). Furthermore, we incorporate a feedback mechanism and adjust the prompt specificity depending on the confidence levels of AI to minimise errors and improve the decision-making capacity (Lin et al., 2023). Contextual framing is used to ensure accuracy. The Practice of Action-oriented Practical Prompt Engineering. This methodology defines and organises data input into AI systems to be clear, complete, and logical, thus providing usable and valuable output.

- **Contextual Framing for Accuracy:** In addition, practical prompt engineering requires some form of contextual framing, ensuring the AI system can access all relevant clinical information for generating a response. In addition, patient-specific details, such as medical history, laboratory tests, or imaging findings, may also be embedded. For instance, a prompt such as "Identify potential causes of shortness of breath in a 40-year-old patient with asthma and recent exposure to environmental allergens" allows the AI system to consider critical clinical parameters to help with more accurate differential diagnoses. Contextual framing helps manage complex cases like multimorbidity, where interdependencies among the conditions need to be considered (Rajpurkar et al., 2022).

2.2 Iterative Testing: Ensuring Reliability and Precision

Further, as part of prompt engineering, iterative testing means repeatedly testing prompts to refine them and have more reliable, relevant AI-generated outputs.« Refinement of Promptsal methodology in prompt engineering is through iteration in Refining Prompts, i.e., repeatedly evaluated to gain higher reliability and relevance of AI-generated outputs.

- **Refining Prompts through Iteration:** In healthcare, where anything can mean life or death, iterative testing helps to avoid mistakes. Initial prompts typically yield refined outputs, which, through numerous iterations of the prompting system, can become optimised for performance. For example, an AI-based triage tool performing a case study had difficulty with the initial prompts owing to ambiguities in the patient's symptoms. However, after several rounds of structured language, plus symptom details, the system's accuracy rose by 30%, lessening emergency department triage misclassifications by a smaller multiple (Smith et al., 2021).

- **Methods for Iterative Refinement:** Two main approaches are employed in iterative testing. Automated feedback loops, including manual evaluation, are used. Healthcare professionals then manually evaluate AI outputs in case they can be improved. At the same time, this is an efficient approach for complex tasks in clinical practice, such as developing treatment recommendations or measuring diagnosis accuracy. On the other hand, automated feedback loops follow predefined metrics such as relevance, accuracy and consistency scores for the refinement process. Not only does it speed up iterative testing, but it also allows real-time adjustments to clinical workflows (Topol, 2019). Iterative testing capabilities are being further improved with emerging technologies. For example, according to newly available patient data, AI systems can automatically adapt prompts into electronic health records (EHRs). This real-time feedback mechanism enables the parallel evolution of prompts with the patient's clinical course, increasing the system's utility in dynamic settings (Lin et al., 2023).

2.3 Incorporating Domain Knowledge: Enhancing Clinical Precision

Prompts include domain knowledge to avoid medical outputs being out of sync with currently accepted medical standards and evidence-based practice. In order to convert the unstructured textual data, we need to use standardised medical terminologies, structured query frameworks, and links to clinical guidelines. The prompts contain embedded medical terminology that ensures that the AI outputs match medical standards and evidence-based practices. Embedding standardised medical terminologies, the ability to perform structured queries on the information and a reference to clinical guidelines are involved.

- **Embedding Medical Terminologies:** How do medical standardised terminologies (International Classification of Diseases, ICD; Systematized Nomenclature of Medicine, SNOMED) allow prompting to reduce the output of potential ambiguity and [accuracy]? In another example, if the radiology-oriented AI system was asked to find abnormalities of a left lung compatible with a pneumothorax, it returned much more precise answers compared to when the same AI system is asked to identify abnormalities on chest Xray (Greenhalgh et al., 2020). When embedded, such terminologies allow AI systems to produce their output in the clinical language and professional standards.

- **Leveraging Structured Query Formats:** Here, the structured query formats, i.e. Population, Intervention, Comparison, and Outcome (PICO), enable the formulation of structured prompts towards answering specified clinical questions. For instance, if a prompt reads, 'Compare the efficacy of ACE inhibitors in reducing hypertension in the elderly with beta-blockers,' the PICO format helps to generate a more specific and comprehensive answer. In addition, this methodology improves the output quality while enforcing evidence-based decision-making by instructing the AI to consider clinically relevant parameters (Goldstein et al., 2023). Linking themselves to evidence-based guidelines means that AI outputs are pioneered in prescribed medical standards and evidence-based practices. This adds embedding standardised medical terminologies, using structured query frameworks and referencing clinical guidelines.
- **Aligning with Evidence-Based Guidelines:** AI results are used as prompts for evidence-based guidelines on how they help clinical best practices. Treatment recommendation prompts can call upon protocols from sources like the American Heart Association and the World Health Organization to increase their reliability. Wang et al. (2020) found that prompts containing explicit reference to these guidelines increased the concordance of AI-generated recommendations with clinicians' decisions by 25% for an AI-driven treatment planning task.

2.4 Integrating LLMs with Existing Infrastructure

Integrating LLMs into healthcare infrastructure means aligning AI capabilities that can be positioned with current electronic health records (EHR) and clinical workflows. A structured approach includes:

- Data Mapping: EHRs have patient records, and prompt designs map these records to LLM inputs for accuracy.
- Real-Time Feedback: Dynamic data streams exist in Wearable and IoT devices for making responsive decisions (Sun, Li, & Wang, 2022).
- Standardisation and Interoperability: Liu et al. (2023). Standards for platform integration of AI-EHR interfaces to ensure a smooth integration.

To automate administrative tasks, LLMs were integrated into an EHR system of a metropolitan hospital. The system can generate concise patient summaries using prompt engineering and flag critical alerts. It led to a 25% reduction in administrative workload so that the physicians could attend to the care of patients (Wang et al., 2023).

2.5 Applications in Healthcare

With prompt engineering methodologies, we have successfully applied them across multiple healthcare domains to enable higher clinical accuracy, efficiency and patient engagement.

- **Differential Diagnoses:** Differential diagnosis generation is particularly successful in these AI systems trained with well-crafted prompts. An example of this comes from a study with infectious disease diagnostics, showing prompts asking for patient demographics, symptom onset, and travel history, resulting in a 40% increase in diagnostic accuracy over generic queries (Kumar, Patel, Ahmad, and Hallbeck, 2022). This is particularly useful in resource-limited settings where early and accurate diagnosis will dramatically alter patient outcomes.
- **Treatment Planning and Precision Medicine:** Including patient-specific information in queries through prompt engineering permits the personalisation of treatment planning. The AI system then generates treatment responses for prompts, such as the following as an example: "Suggest treatment options for a 65-year-old male with advanced prostate cancer and comorbidity diabetes," which are tailored to oncologic guidelines. Specifically, this is useful in precision medicine (Lee et al., 2021), an area where, more precisely, treatment decisions ought to be contemplated if they do not always incorporate hereditary, actual, and way-of-life elements.
- **Patient Education and Patient Empowerment:** Prompt engineering also simplifies medical conditions or treatment explanations for patient education. If I were to prompt questions regarding the most common patient concerns, such as 'Elaborate on the symptoms, management and long term outlook of Type 2 diabetes in plain language,' the patients understood and engaged much better. It is found that such approaches result in increases in adherence to treatment plans and informed decision-making (Wang et al., 2020).

2.6 Challenges and Future Directions

The promise is auspicious, but there are a few challenges with prompt engineering in healthcare, e.g., aspects of ethics, biases in AI systems, and the input forms having to be overwritten on a schedule corresponding to new medical knowledge. Ethical and Privacy Concerns, also. Many challenges exist in healthcare engineering, ranging from ethical factors to bias in AI systems to ever-changing medical knowledge.

- **Ethical and privacy concerns:** Okay, but what will you do with prompts, and how do you take care of them without compromising patient privacy and line up with things like the Health Insurance Portability and Accountability Act (HIPAA)? Prompts should not incidentally impart personal patient information while maintaining clinical usefulness (Johnson et al., 2023).
- **Mitigating Biases in AI:** It turns out that AI systems are highly susceptible to biases because they are trained on imbalanced training datasets or poorly constructed prompts. To address these biases, we need diverse and representative datasets and rigorous fairness testing across demographic groups (Buolamwini & Gebru, 2018).
- **Dynamic Medical Knowledge:** Engineering methodologies must be updated quickly because medical innovation is fast-paced. Future work will focus on adaptive systems integrating new evidence, guidelines and clinical insight in real time. These advancements will enable AI to become a necessary and relevant tool in the ever-changing environments of healthcare research (Smith et al., 2023). To ensure that AI is successfully integrated into healthcare, prompt engineering must be performed to produce accurate, context-sensitive, clinically relevant outputs. Iteratively testing and incorporating domain knowledge with OK tuning prompts, AI has the chance to increase diagnostic accuracy, treatment planning and patient education efforts.

Table 3. Ethical Challenges in Prompt Engineering

S.No.	Challenge	Description	Potential Solutions
1	Data Privacy	Risk of exposing sensitive patient data	Implement anonymisation techniques
2	Bias in AI Outputs	Inherent biases in training datasets	Use diverse and representative datasets
3	Overdependence on Technology	Automation bias in clinical decisions	Ensure human oversight in decision-making

3. APPLICATIONS OF PROMPT ENGINEERING IN HEALTHCARE: CASE STUDIES

Artificial intelligence (AI) integration has completely transformed the field of healthcare. Prompt engineering has been characterised as a revolutionary approach to optimising healthcare delivery. By narrowing how AI systems perceive and react to user input, prompt engineering can help increase diagnostic accuracy, strengthen patient-provider dialogue, and improve healthcare efficiencies. The emergence of these applications is shown through case studies with recent advancements in the field.

3.1 Improving Diagnostic Accuracy

There can be no effective patient care without accurate diagnosis, and misdiagnosis remains a significant problem, with sizeable preventable morbidity and mortality. Using prompt engineering, AI diagnostic tools can generate a complete differential diagnosis with the most minor information available: the patient's symptoms, clinical data, and relevant background information. For example, rare diseases can be diagnosed with better outcomes from AI-based AI systems when prescribed to include prompts for rare diseases. In a recent study, AI model diagnostic error rates on atypical presentations were 15% lower than conventional algorithms using engineered prompts (Jones et al., 2023). Also, prompts created by AI technologies can facilitate time-sensitive diagnoses. On the other hand, prompt engineering in emergency settings prevents AI systems from waiting until other conditions, like acute myocardial infarction, are elevated. They would rather grade symptoms as high-risk conditions. For instance, prompt enhanced AI proved practical by cutting down critical case time to diagnosis by 20% following a real-world deployment in a tertiary care emergency department (Smith et al., 2023). Effectively, prompt engineering can assemble and coordinate others, like lab results, imaging, and genetic data—this benefits the diagnostic utility of prompt engineering. These systems dynamically change prompts corresponding to revised clinical input, supplying real-time diagnostic support to healthcare providers, enabling them to refine decision-making and improve patient outcomes.

3.2 Enhancing Patient-Provider Communication

Communication is key to patient-centred care, but this communication, marked with medical jargon and otherwise complex information, could be more comprehensible to patients, especially those with low health literacy. The challenge is met by prompt engineering, which creates personalised and simplified medical information for patients to engage with. In a recent randomised trial, we examined the effect of tailored AI-generated prompts on patient comprehension during oncology consultations. Patients who received information via AI-generated prompts were found to have an improved understanding of treatment options by 30 per cent compared to a control group (Garcia et al., 2023). These were prompts intended to reframe complex terms into user-friendly language without sacrificing accuracy. Moreover, prompt engineering enables decision-making to be shared. Using research revealing that prompts designed to elicit patient preferences can guide treatment trade-off discussions, Finch et al. developed a novel clinical decision aid to help individuals make treatment decisions. An example is a case study in which patients with chronic kidney disease discussed their values related to dialysis versus conservative care using AI-generated prompts. This approach helped build trust and increased compliance with agreed-upon treatment plans (Kumar et al., 2023). Aside from enhancing patient understanding, prompts also fill the cultural and linguistic gap. Prompt engineering for multilingual settings gives us confidence that an AI system can deliver the correct translation and a culturally appropriate interpretation necessary to support communications across diverse patient populations. In particular, this application has been critical in rural and underserved areas where language barriers increase health inequities (Chen et al., 2023). Prompt engineering enables the provider to empower patients with tailored information to consume and facilitate dialogue, improving individual patient health outcomes and the provider-patient relationship.

3.3 Streamlining Workflow and Reducing Burnout

Typically, clinician burnout results from heavy administrative burden and decreases job satisfaction and patient care quality. The solution is prompt engineering, automating repetitive tasks so clinicians can focus on what they do best – direct patient care. Up to 40% of a clinician's workday involves administrative tasks, including charting, documentation and coding. Well-designed prompts can help AI systems simplify these tasks. As an example, a study on 500 primary care physicians found that adding prompt engineering to EHRs also decreased time spent on documentation by 25%, with 90% of participants noting better workflow efficiency (Wang et al., 2023). Another critical application involves clinical decision support systems (CDSS). CDSSs are engineered through prompt engineering to provide actionable insights while minimising noisy or redundant information for clinicians. To that end, alerts in a multicenter trial, for example, with engineered prompts to CDSS, reduced alert fatigue, possibly a common source of burnout, by filtering out noncritical notifications. By 40% less ignored alerts, this intervention improved clinician response to critical cases (Lee et al., 2023). Nevertheless, besides administration, prompting also helps in scheduling or resource allocation. Healthcare systems use AI-driven prompts to analyse appointment data and determine the expected no-show rates to adjust their schedule dynamically. In addition, we tested prompt-based scheduling in a case study from a metropolitan hospital and showed that it reduced missed appointments by 18%, thereby improving operational efficiency (Brown et al., 2023). Ultimately, 'prompt engineering' benefits clinicians' psychological well-being; thus, it reduces the cognitive load by automating meaningless tasks and providing intuitive decision support, supporting a more sustainable working environment. This retention of skilled healthcare professionals comes at the expense of the system combating burnout.

4. BROADER IMPLICATIONS OF PROMPT ENGINEERING IN MEDICAL EDUCATION AND MULTIDISCIPLINARY COMMUNICATION

The introduction of artificial intelligence (AI) has made prompt engineering the most important method for creating and using artificial intelligence substantial language models (LLMs). From this arises the careful development of prompts to incentivise the generation of correct AI responses while ensuring their relevance in context. Prompt engineering has implications beyond computational efficiency, especially in use cases of related domains, such as healthcare. Prompt engineering fundamentally changes medical education practices and interprofessional communication by creating a space for this novel approach to critical thinking, decision making and collaboration.

4.1 Enhancing Simulations with AI-Driven Prompt Engineering

Simulations have been an important component of medical education over the long term because they enable the trainees to train in a risk-free environment where they can practice procedures, make decisions, and diagnose conditions. AI-driven simulations have become far better learning tools, and their application has become far broader with the help of prompt engineering. We show how LLMs can replicate patient interactions using carefully crafted prompts to mirror the dynamic factors that affect real-world patient problems and react to those nuanced ones. These AI-powered tools have since been validated in medical training and have had positive results. For example, Lin et al. (2023) studied that prompt-engineered LLMs significantly boosted diagnostic accuracy in a cohort of medical students. The participants were presented with AI-driven patient scenarios with differential diagnoses and treatment planning. These sessions used precise, interactive prompts, facilitating critical thinking and adaptability. In addition, Zhang et al. (2022) demonstrated that prompt-engineered AI could support improving surgical training by conducting virtual simulations of rare surgical cases with feedback tailored to decisions made during surgery. Moreover, learning anything in a personalised fashion is made possible through prompt engineering. This is valuable since the adaptability of these systems is critical for sustaining learner motivation and incremental skill acquisition (Smith et al., 2023).

4.2 Fostering Critical Thinking and Decision-Making

Critical thinking and decision-making are foundational competencies of a medical professional. The complex, multifaceted scenarios AI can now prompt trainees to test these skills would only be possible with prompt engineering. Prompts of ethical dilemmas in patient care provide an example; these prompt learners to consider the trading off of priorities like patient autonomy against medical best practices. According to a controlled study by Ahmed et al. (2023), medical residents trained with AI-powered ethical scenarios improved their ability to articulate and justify their decisions by 23%, compared to traditional case-based methods. Additionally, prompt-engineered AI can be used for real-time evaluation of trainee responses, which can be used to provide instant feedback and reinforce learning. Being iterative, these interactions enable trainees to refine their approaches with AI-provided insights (Miller & Green, 2023). This iterative feedback loop is very beneficial in developing skills like risk assessment, resource allocation, and patient communication.

4.3 Addressing Barriers to Traditional Training

These barriers in traditional medical training are also dealt with by prompt engineering. Trainees in such areas need more access to diverse clinical cases. Today, the gap in real-world exposure is filled by AI-driven simulations, fed with powerful prompts that give us virtual-world exposure to every possible condition and situation. Medical students in underserved regions who used AI simulations found a 40 per cent improvement in their confidence levels when encountering unfamiliar conditions during their clinical rotations, as evidenced by Chen et al. (2023).

4.4 Standardizing Medical Language

Communication in multidisciplinary healthcare settings must be adequate, where various professionals come from different specialities to work as one on patient care. Standardising medical language is what prompt engineering does to the medical language to have clarity and consistency in the interactions facilitated by the AI systems. LLMs are prompted with structured prompts to yield clear, context-aware outputs and reduce ambiguity or misinterpretations. For example, structured prompts can tell an AI system to summarise patient histories in standard formats so doctors can communicate seamlessly. However, recent research shows how prompt engineering can open that door. LLMs were deployed and prompt-engineered to improve communication and coordination in multidisciplinary team meetings to facilitate communication among physicians, nurses, and social workers (Taylor et al., 2023). Perspectives were aligned by structuring prompts that helped the AI system provide a summary and action items of key points so all team members understood the treatment plan. According to the participants, the prompt engineering of the conversations bridged the linguistic and disciplinary gaps to yield a 27% improvement in perceived communication efficiency.

4.5 Improving Coordination in Patient Management

Managing patients usually means complex coordination between several stakeholders involved in patient care, such as physicians, nurses, pharmacists and allied health professionals. This collaboration process is further enhanced by prompt engineering, allowing AI systems to mediate information exchange and task delegation. For example, prompts can tell an AI system to spot potential conflicts in medication regimens given to patients by different specialists so they can be flagged for review. Patel et al. (2023) perform a clinical trial in which the AI system, with its prompt engineering capability, is plugged into a hospital's electronic health record (EHR) system. The system enabled real-time communication between team attendees, updating patient status, medication changes, and test results. However, coordination via AI that resulted from this intervention led to a 15% decrease in adverse drug interactions and a 20% improvement in patient outcomes—in tangible benefits.

4.6 Enhancing Inclusivity and Equity

Additionally, prompt engineering is conducive to broader healthcare inclusivity and equity communication. By creating prompts that factor in cultural and linguistic diversity, AI systems can produce outputs that will be meaningful to different populations of patients and the healthcare providers who are working with them. For example, prompts can be set for AI systems to translate medical jargon into a non-specialist's term for patient education. This ability is beneficial when communication effectively enables equitable care delivery (Nguyen et al., 2023). Even if it shows potential, integrating prompt engineering into medical education and interprofessional communication is not without problems. The existence of biases within prompts is one concern since if these issues get embedded into AI outputs, then these propagate those sorts of inequities throughout training and care. This need to solve this problem calls for rigorous testing and validation of prompts to ensure exclusive and representative prompts for different populations (Kumar et al., 2023). Longitudinal studies of prompt engineering in medical education and patient outcomes with extended follow-up are also necessary. Additionally, further exploring the possibility of integrating prompt-engineered AI with new technologies, such as virtual reality and wearable devices, can increase its usefulness. Moreover, prompting in healthcare should significantly democratise access to AI-based tools in low-resource settings. Prompt engineering is an innovative model of using AI in medical education and other disciplines. With this technique, medical language is standardised and interactive, and personalised simulations are provided, strategically increasing critical thinking, decision-making, and collaboration in healthcare. Its efficacy has evidence to back it up, overcoming traditional barriers in training, enhancing patient management and being inclusive at the same time. Nevertheless, bias, ethics, and scaling remain central challenges prior to their ability to sustain impact. Given the presence of prompt engineering in healthcare today, along with the evolving technology used in data processing of medical data, an efficient, equitable and patient-centric healthcare system is predicted to be the case soon, if not already.

5. CHALLENGES AND ETHICAL CONSIDERATIONS IN PROMPT ENGINEERING FOR HEALTHCARE

Large language models (LLMs) and prompt engineering have been siloed, but the realisation is that some of the applications of LLMs to healthcare are changing the game. This allows specific, contextual queries to be crafted tailored to the outputs of LLMs in tasks such as diagnostics and personalised treatment planning. While it offers tremendous promise, integrating LLMs into healthcare is dogged by practical and ethical issues that need careful thinking and more proactive management. In this chapter, the authors discuss three main ethical issues: data privacy, bias in AI outputs, and technology dependency, emphasising the need to develop evolving ethical frameworks for tackling the challenges in those areas.

5.1 Data Privacy: Protecting Confidentiality in LLM Training and Deployment

However, using patient data to train LLMs has grave implications for confidentiality and data security. Sensitive patient health information (PHI) is handled in a data-handling process that can cause serious consequences such as identity theft, discrimination and a general loss of faith in medical care processes (Sweeney, 2015). Like other LLMs, the current training methods create significant data aggregation, increasing the chance of exposure to the data.

5.1.1 Advances and Challenges in Data Protection

These concerns have been addressed in recent work using federated learning and differential privacy. Using Federated learning to train LLMs on decentralised data offers the advantage of training on sensitive data, which stays decentralised on local devices (Li et al., 2020). Dwork and Roth (2014) explain that differential privacy techniques inject statistical noise in datasets to ensure a minimum risk of re-identification and a maximum data utility for training. Noise addition will reduce model performance, but federated learning will still be vulnerable to sophisticated inference attacks (Hitaj et al., 2017). Another part of the risk is the benefit of using existing cloud-based services to deploy the model. These platforms often spread across jurisdictions, leading to disagreements about what data to protect and how to do so under data protection laws such as the European Union's GDPR or US HIPAA. The challenge is to align with these frameworks while not reducing the global utility of LLMs, which is a difficult task (Voigt and von dem Bussche, 2017).

5.1.2 Ethical Imperatives for Data Privacy

To tackle these problems, we need to prioritise robust consent mechanisms and transparent data governance practices. Patients should have the power to control how their data is utilised and be able to opt themselves out of the data used in AI training processes. Moreover, healthcare organisations need to invest money into cybersecurity infrastructure to diminish the dangers of unapproved data access and breaks. Finally, regular audits and updates of privacy-preserving methods are also required as a matter of ethical guidelines to match the pace of fast-developing threats.

5.2 Bias in AI Outputs: Ensuring Diagnostic Accuracy and Fairness

AI systems in healthcare carry significant risks of bias, which can arise from real-world datasets, the design and implementation of training processes, or the inherent structure of models themselves. These biases can propagate through Large Language Models (LLMs), amplifying disparities and leading to unequal treatment recommendations or diagnostic errors. This issue is particularly concerning in healthcare, where such biases can have life-or-death consequences. For example, historical inequities embedded in training data may cause LLMs to underperform for underrepresented populations, resulting in misdiagnoses or inappropriate treatment plans. Addressing these challenges requires rigorous validation processes, inclusive datasets, and algorithms designed to detect and mitigate bias effectively, ensuring equitable and accurate AI-driven healthcare outcomes.

5.2.1 Sources and Manifestations of Bias

There are many ways AI outputs can become biased: due to historical inequities in the healthcare system, imbalanced datasets, and exceptional algorithmic design choices. For instance, a current study (Adamson & Smith, 2018) revealed that AI models trained on almost entirely Caucasian studies reduced accuracy in diagnosing skin problems in patients with darker skin tones. It explains why inclusive and representative datasets are needed. Moreover, the interpretiveness of LLMs can make bias worse. Unlike structured algorithms, LLMs create output based on probabilistic associations that might inadvertently mirror the stereotypes of society that are present in training data (Bolukbasi et al., 2016). This could appear as gendered or racial diagnostic suggestions or treatment recommendations in healthcare.

5.2.2 Strategies to Mitigate Bias

In order to reduce bias, training datasets need to be diversified to capture the entire lineup of patient demographics. We finally describe how synthetic data can be generated to cover the holes of the unrepresented demographic sets, which has been proven as an alternative (Frid-Adar et al., 2018). Additionally, I wish to see bias detection and mitigation algorithms enter the pipeline to identify and undo the skewness of the outputs. An additional encouraging direction focused on leveraging XAI techniques to learn and assess how clinicians make decisions related to LLMs (Rudin, 2019).

5.2.3 Ethical Considerations for Bias Management

The ethical guidelines to be covered by AI development and deployment must deal with accountability. We should require developers to document their steps to sidestep bias as part of their report. Regulatory bodies must also set standardised benchmarks to which healthcare applications assess consistency in detecting and mitigating bias across applications. These standards will be critical when establishing them for collaboration between technologists, clinicians, and ethicists.

5.3 Dependency on Technology: Balancing AI and Clinical Judgment

As AI infiltration into healthcare continues, there is a growing worry that reliance on the technology will erode clinical judgement and overdependence will ensue. LLMs are undoubtedly helpful, but they make neither good nor bad decisions; they are merely tools to be used in service of a human, to augment, not replace, human expertise.

5.3.1 Risks of Overdependence

A significant risk is 'automation bias', in which clinicians depend too heavily upon AI recommendations, even if they do not coincide with clinical intuition or evidence (Skitka et al., 1999). Regarding high-stakes cases, such as emergency care or rare disease diagnosis, LLMs can be incredibly dangerous because they lack training data to make adequate recommendations. Another worry is that many LLMs are "black boxes," meaning we cannot easily understand why they produce the output they do. The consequence of this lack of transparency is blind trust in AI systems, such that crucial thinking skills of healthcare providers are affected (Zhang et al., 2021).

5.3.2 Fostering a Balanced Approach

These risks can, however, be mitigated with the model of AI as a support tool rather than as a decision-maker by healthcare institutions. Clinicians must learn new skills, which should be provided through the updated training programs, to evaluate AI outputs and integrate them into clinical practice critically. All of this should be accompanied by teaching people to understand the limitations and risks of LLMs. In addition, regulatory frameworks should mandate a 'human-in-the-loop' approach, which leaves clinicians with final decision-making over medical aspects. The latter can be achieved via safeguards, such as the requirement for manual confirmation for high-risk recommendations or second opinion mechanisms.

5.3.3 Ethical Guidelines for Dependency Management

Regardless, frameworks for deriving ethical decisions must privilege human agency in healthcare. Clear boundaries around the application of AI and protocols for resolving conflicts between AI recommendation and clinical judgment were included. Healthcare uses, without a doubt, the most attractive cases for prompt engineering. However, difficulties with ethical and practical implications remain the roadblocks of its use in a clinical setting. For data privacy, the outputs of AI are biased towards certain groups or dependent upon technology, so we need to take a holistic approach to improved technical innovation supplemented by sound ethical oversight. As AI progresses, so, too, must the principles and frameworks that underpin AI-powered tools used in medicine so that these tools are used in ways that adhere to the highest standards of fairness, accountability, and patient care.

5.3.4 The Role of Prompt Engineering in Transforming Healthcare

Artificial intelligence (AI), healthcare, large language models (LLMs), and prompt engineering have recently created a paradigm shift. They are revolutionising how personalised, efficient, and scalable care is delivered. Therefore, it is critical to examine and tackle issues and prospects of the future of the field as it evolves to realise AI's transformative power responsibly and efficiently.

6. FUTURE DIRECTIONS

- Real-Time Feedback Integration: Using wearable device data and AI to help create dynamic, patient-specific insights.
- Multilingual LLMs: Developing AI models to increase healthcare access in non-English speaking regions.
- Predictive Analytics: AI-powered prompts help improve risk stratification by utilising genetic and environmental data.

6.1 Real-Time Patient Feedback as Applied to Prompt Design

Making patient feedback real-time while integrating it into prompt engineering serves as a frontier to improve healthcare outcomes. On the other hand, patient-generated data (e.g., symptoms, preferences, and how they have responded to treatment) can help AI models provide more personalised recommendations. For example, now we can constantly monitor physical parameters with the help of sensitive wearable technology and mobile health applications. Integrating these data streams into LLMs enhances the accuracy of prompts for making diagnostic and putative therapeutic decisions (Sun et al., 2022). An approach that works is shared decision-making augmented with honest time feedback, resulting in improved patient satisfaction and adherence to care plans. According to Mahajan et al. (2023), integrating patient-reported outcomes within an AI-driven decision support system reduced diagnostic errors in primary care by 15%. However, real-time integration will only achieve its maximum potential as future research overcomes the challenge of harmonising heterogeneous data sources and ensuring data quality and patient privacy.

6.2 Exploring the Role of Multilingual LLMs in Global Healthcare

After all, equitable healthcare delivery worldwide is a disastrous dream, even with language barriers. We can fill these gaps using multiple languages with prompt engineering and multilingual LLMs. GPT4 and Bloom are trained in many languages and are usable for multilingual applications (Brown et al., 2023). Multilingual LLMs enable accurate translations of medical consultations, patient education materials and dissemination of research in the languages underrepresented in literature, for example. However, translations by such a Swahili-adapted LLM, used in a recent pilot study in sub-Saharan Africa (Mwangi et al., 2023), suggest that such translations can improve the accessibility of healthcare information to rural populations by 38%. Also, LLMs can help healthcare professionals from low-resource areas rapidly access medical knowledge in their language. Future efforts will be to reduce biases in multilingual models and enlarge the dataset with low-resource languages.

6.3 Advancing Ethical AI Frameworks Tailored to Healthcare

In healthcare, more than in any other sphere, the introduction of AI is a veritable problem with ethical dimensions ranging from algorithmic bias and data privacy to accountability. The vital process in embedding ethical considerations into AI systems is prompt engineering. For example, prompt engineering can achieve equitable outcomes—for example, by flagging bias in diagnostic algorithms or requiring diverse representation in training data sets (Rajkomar et al., 2022). Furthermore, LLM's decision-making should be transparent so that the patients and providers can believe it. Combining ethical, prompt engineering with explainable AI (XAI) frameworks allows AI-generated recommendations to be explained by their rationale. In a study, Yang et al. (2023) highlighted how transparent AI can trigger 24% of clinician acceptance of a decision support tool.

6.4 Enhancing Clinical Workflow Integration

For AI to be helpful in healthcare, it must fit nicely into clinical workflows. In future directions for prompt engineering, adaptive prompts will be designed to suit clinicians' needs, allow the avoidance of cognitive overload, and maximise efficiency. For that, some context-aware prompts that consider a physician's speciality, the demographics of a patient and the history of interaction with a patient can help with documentation, avoid repetition and improve productivity. Regarding the customisation of prompts, electronic health records (EHRs) can be used to increase relevance and accuracy. A study by Liu et al. (2023) showed that customised prompts in a tertiary care hospital reduced documentation errors by 18%. Nevertheless, obstacles to interoperability between EHR programs and AI devices have primarily stalled these solutions. Solving these technical hurdles will be the work of stakeholders, including healthcare institutions, AI developers, and policymakers.

6.5 Expanding the Scope of Predictive Analytics

Both LLMs-powered predictive analytics can transform preventive care and risk stratification. By incorporating this diversity, using genomics, social determinants of health, and environmental factors, for example, prompt engineering can boost the predictive ability of AI. Prompts to analyse genomic data can check individuals for hereditary conditions for which they may be at risk so necessary steps can be taken earlier. Furthermore, population-level data may be incorporated to assist with predicting and controlling outbreaks. During the time of the COVID-19 pandemic, prompt-engineered LLMs helped to predict infection trends and resource allocations (Wang et al., 2023). Future work should also investigate ways to combine forthcoming data types (e.g. microbiome profiles and epigenetic markers) into predictive analytics packages. In healthcare, prompt engineering is poised to drive innovation in diagnosis, treatment and patient engagement. Prompt engineering can close existing disparities and open new ways to improve global healthcare by integrating real-time feedback in prompt engineering, advancing multilingual capabilities, and embedding ethical considerations. However, researchers, clinicians, and policymakers need to make a significant effort to realise this potential. Among these challenges, standardising the data, algorithmic bias, and interoperability stand out as morally responsible and equitable in using AI in medicine. However, given that LLM advancements have recently accelerated — as Lasso puts it —prompt engineering will sit at the crux of how to unlock the potential of AI itself within digital healthcare. The intersection of human expertise and machine intelligence is the future of healthcare. By closing the gap, prompt engineering has great potential to help improve clinical outcomes, decrease errors, and empower patients.

Table 4. Benefits of Prompt Engineering in Clinical Workflows

S.No.	Area	Impact	Example
1	Administrative Tasks	Reduced time spent on documentation	AI-assisted EHR documentation
2	Decision Support	Improved accuracy of clinical decisions	Prompt-engineered CDSS alerts
3	Scheduling	Optimised resource allocation	Dynamic appointment scheduling

7. CONCLUSION

Prompt engineering, which closes the gap between what health care can do with AI and what you need it to do, aids in improving accuracy in diagnosing, patient education, and operational efficiency. Challenges such as bias and privacy are addressed to achieve AI's equal integration. Prompt engineering will redefine healthcare by equipping clinicians with tools and delivering scalable, patient-inclusive solutions as we continue to innovate. Prompt engineering is integrated within healthcare to solve critical problems and increase operational efficiencies. Using the methodology, generating well-structured inputs for precise interaction with AI to improve diagnostic accuracy, patient communication, and workflow is possible. It allows clinicians to tailor the output to context-specific scenarios, engenders shared decision-making, and supports equitable care delivery. Additionally, AI implants its way into medical education and multidisciplinary communication, facilitating better development of critical thinking skills, standardising medical language, and improving communication between healthcare teams. Prompt engineering can be used for tasks not initially designed for, allowing the realisation of its full potential but highlighting ethical concerns that need to be addressed to realise this potential, including minimising bias in models, balancing technological reliance with clinical judgment, and protecting data privacy. Improvements to future systems include fine-tuning multilingual abilities and incorporating real-time patient feedback and ethical AI frameworks to guarantee inclusiveness and fairness. As healthcare systems grow, prompt engineering represents a key innovation that marries human expertise with machine intelligence to deliver scalable, patient-centric services.

REFERENCES

Adamson, A. S., & Smith, A. (2018). Machine learning and health care disparities: A diagnostic algorithm learns implicit bias. *JAMA Dermatology*, *154*(11), 1247–1248. DOI: 10.1001/jamadermatol.2018.2348 PMID: 30073260

Ahmed, S.. (2023). Ethical decision-making training using AI-driven simulations: A comparative study. *Medical Ethics Quarterly*, *35*(2), 154–170.

Beauchamp, T. L., & Childress, J. F. (2013). *Principles of Biomedical Ethics*. Oxford University Press.

Bolukbasi, T., Chang, K. W., Zou, J. Y., Saligrama, V., & Kalai, A. T. (2016). Man is to computer programmer as woman is to homemaker? Debiasing word embeddings. *Advances in Neural Information Processing Systems*, •••, 29.

Brown, T.. (2020). Language models are few-shot learners. *Advances in Neural Information Processing Systems*, *33*, 1877–1901.

Brown, T., Mann, B., & Ryder, N.. (2023). Scaling language models for global applications: Insights and challenges. *Journal of Artificial Intelligence Research*, *67*(4), 321–340. DOI: 10.1016/j.jair.2023.04.003

Buolamwini, J., & Gebru, T. (2018). Gender shades: Intersectional accuracy disparities in commercial gender classification. *Proceedings of the Conference on Fairness, Accountability, and Transparency*, 77-91.

Chen, Y.. (2023). Bridging training gaps in underserved regions with AI simulations. *Journal of Medical Education*, *48*(3), 212–228.

Chen, Y., Zhang, X., & Lee, R. (2023). Enhancing multilingual patient communication using AI-driven prompt engineering: A rural health case study. *Journal of Medical Informatics*, *45*(3), 243–259.

Dwork, C., & Roth, A. (2014). The algorithmic foundations of differential privacy. *Foundations and Trends in Theoretical Computer Science*, *9*(3–4), 211–407.

Frid-Adar, M.. (2018). Synthetic data augmentation for improved liver lesion classification in CT. *IEEE Transactions on Medical Imaging*, *37*(3), 759–769.

Garcia, L., Patel, S., & Anderson, B. (2023). Simplifying complex medical information through AI-generated prompts: A randomised trial in oncology. *Cancer Communications*, *38*(2), 102–115.

Goldstein, D.. (2023). Guidelines-based prompts in AI treatment recommendations. *Journal of Clinical AI, 10*(2), 145–152.

Greenhalgh, T.. (2020). Evidence-based medicine: A guide. *BMJ (Clinical Research Ed.), 371*, m4425.

Hitaj, B.. (2017). Deep models under the GAN: Information leakage from collaborative deep learning. *Proceedings of the ACM SIGSAC Conference on Computer and Communications Security*. DOI: 10.1145/3133956.3134012

Jiang, H.. (2022). Enhancing context in AI healthcare applications. *Journal of Biomedical Informatics, 128*, 104010.

Johnson, A.. (2023). Ethical AI in healthcare. *Journal of AI Ethics, 4*(1), 25–40.

Johnson, P., Lee, A., & Smith, R. (2023). Reducing Diagnostic Errors Using AI-Powered Decision Support Systems. *Journal of Medical Systems, 47*(4), 45.

Johnson, S.. (2023). AI in Medicine: Leveraging Large Language Models for Clinical Excellence. *Journal of Medical AI Research, 12*(3), 101–115.

Jones, D., Miller, T., & Hargrove, K. (2023). Leveraging AI for rare disease diagnosis: The role of prompt engineering in improving accuracy. *The Lancet. Digital Health, 5*(6), e234–e241.

Kumar, P.. (2023). Addressing bias in prompt engineering for healthcare AI applications. *AI Ethics Journal, 15*(1), 45–60.

Kumar, R., Singh, A., & Banerjee, P. (2023). Tailored AI prompts for shared decision-making in chronic disease management. *Journal of General Internal Medicine, 38*(4), 876–883.

Kumar, S.. (2022). Differential diagnosis generation in AI systems. *AI in Medicine, 14*(4), 233–242.

Lee, A., & Smith, J. (2023). Patient Engagement through AI-Driven Communication Tools. *Digital Health Review, 8*(2), 45–60.

Lee, J., Park, H., & Wang, S. (2023). Reducing alert fatigue in clinical decision support systems with prompt engineering. *Health Informatics Journal, 29*(1), 15–28.

Lee, M.. (2021). Personalised treatment planning with AI. *Journal of Medical AI Applications, 9*(2), 89–95.

Li, T., Sahu, A. K., Talwalkar, A., & Smith, V. (2020). Federated learning: Challenges, methods, and future directions. *IEEE Signal Processing Magazine, 37*(3), 50–60. DOI: 10.1109/MSP.2020.2975749

Lin, J.. (2023). Improving diagnostic accuracy with prompt-engineered AI tools in medical education. *International Journal of Medical Informatics*, *129*(4), 97–110.

Lin, Z.. (2023). Integrating electronic health record data into AI systems. *Clinical AI Research*, *12*(1), 45–56.

Liu, Y., Zhang, W., & Chen, L.. (2023). Context-aware prompt engineering for reducing documentation errors in clinical settings. *Health Informatics Journal*, *29*(3), 567–581. DOI: 10.1177/146045822311233

Mahajan, V., Singh, A., & Sharma, R. (2023). Enhancing diagnostic accuracy with patient-reported outcomes: A real-world study. *BMJ Open*, *13*(5), e056789. DOI: 10.1136/bmjopen-2023-056789

Mehrabi, N., Morstatter, F., Saxena, N., Lerman, K., & Galstyan, A. (2021). A Survey on Bias and Fairness in Machine Learning. *ACM Computing Surveys*, *54*(6), 1–35. DOI: 10.1145/3457607

Miller, A., & Green, R. (2023). Real-time feedback in medical training: The role of prompt engineering. *Advances in Health Education*, *12*(1), 45–67.

Mwangi, J., Njoroge, P., & Otieno, J. (2023). Application of multilingual language models in rural healthcare: A case study from sub-Saharan Africa. *Global Health Action*, *16*(1), 2103942. DOI: 10.1080/16549716.2023.2103942

Nguyen, L.. (2023). Cultural sensitivity in AI-driven healthcare communication. *Health Equity Reports*, *7*(2), 34–50.

Parmar, H., & Murari, U. K. (2024). Metaverse's significance for smart cities and the retail sector: Facilitating technologies and upcoming approaches. In *Omnichannel approach to co-creating customer experiences through metaverse platforms* (pp. 1–15). IGI Global., DOI: 10.4018/979-8-3693-1866-9.ch009

Patel, R.. (2023). AI-powered coordination in patient management: A clinical trial. *Journal of Health Systems Integration*, *10*(3), 89–105.

Rajkomar, A., Hardt, M., & Howell, M.. (2022). Addressing algorithmic bias in healthcare: Ethical considerations for AI deployment. *Nature Medicine*, *28*(12), 2401–2410. DOI: 10.1038/s41591-022-02018-3

Rajpurkar, P.. (2022). AI reliability through iterative testing. *Nature Digital Medicine*, *5*, 21–32.

Rudin, C. (2019). Stop explaining black-box machine learning models for high-stakes decisions and use interpretable models instead. *Nature Machine Intelligence*, *1*(5), 206–215. DOI: 10.1038/s42256-019-0048-x PMID: 35603010

Smith, A., Brown, C., & Rivera, P. (2023). AI-assisted diagnostics in emergency care: A case study on the impact of prompt engineering. *Emergency Medicine Journal : EMJ*, *40*(1), 12–20. PMID: 36202623

Smith, H.. (2023). Personalized learning with AI: The impact of prompt engineering. *Educational Innovations in Medicine*, *14*(1), 25–39.

Smith, J.. (2021). Iterative improvement in AI triage systems. *Healthcare AI Journal*, *8*(3), 101–112.

Sun, Q., Li, Z., & Wang, H. (2022). Real-time patient monitoring and AI-driven prompt engineering in personalised medicine. *Journal of Biomedical Informatics*, *136*, 104267. DOI: 10.1016/j.jbi.2022.104267

Sweeney, L. (2015). Discrimination in online ad delivery. *Communications of the ACM*, *56*(5), 44–54. DOI: 10.1145/2447976.2447990

Taylor, K.. (2023). Facilitating multidisciplinary communication with AI: A pilot study. *Journal of Clinical Collaboration*, *18*(4), 156–170.

Topol, E. (2019). *Deep Medicine: How Artificial Intelligence Can Make Healthcare Human Again*. Basic Books.

Voigt, P., & von dem Bussche, A. (2017). *The EU General Data Protection Regulation (GDPR): A practical guide*. Springer. DOI: 10.1007/978-3-319-57959-7

Wang, L., Kim, J., & Carter, E. (2023). Streamlining clinical workflows: The impact of AI prompt engineering on electronic health records. *Computational Health Research*, *18*(2), 101–112. DOI: 10.1016/j.chr.2023.05.001

Wang, X., Zhou, Y., & Liu, J. (2023). LLMs and predictive analytics: A pandemic case study. *Computational Health Research*, *18*(2), 101–112. DOI: 10.1016/j.chr.2023.05.001

Wang, Y.. (2020). AI for patient education. *Patient Education Today*, *3*(2), 77–86.

Yang, F., Chen, Y., & Luo, H. (2023). Explainable AI in healthcare: Impacts on clinician trust and adoption. *AI in Medicine*, *142*, 102356. DOI: 10.1016/j.artmed.2023.102356

Zhang, W.. (2022). Advancing surgical training with AI-powered simulations. *Surgical Education Quarterly*, *29*(2), 87–102.

Zhang, Y.. (2021). Black-box explanations in AI: Towards human-centric AI systems. *Journal of Artificial Intelligence Research*, *70*, 701–743.

Chapter 7
The Fuzzy Frontier: Exploring the Potential of Uncertainty in Healthcare

Santhosh Kumar Rajamani
https://orcid.org/0000-0001-6552-5578
MAEER MIT Pune's MIMER Medical College, India & Dr. BSTR Hospital, India

Radha Srinivasan Iyer
https://orcid.org/0000-0001-7387-4401
SEC Centre for Independent Living, India

ABSTRACT

The integration of fuzzy logic and related approaches in healthcare has emerged as a promising avenue for addressing the inherent complexities and uncertainties in medical data and decision-making processes. This review chapter explores the application of fuzzy logic in various domains of healthcare, including medical diagnosis, treatment planning, patient monitoring, and healthcare management. By accommodating imprecise information, incomplete data, and subjective interpretations, fuzzy approaches offer a flexible framework for modelling and analyzing healthcare systems. The chapter examines the principles of fuzzy logic, its implementation in healthcare contexts, and the potential benefits it brings, such as enhanced decision support, personalized treatment strategies, and improved patient outcomes. Furthermore, challenges and limitations associated with the adoption of fuzzy approaches in healthcare are discussed, along with future research directions. This chapter serves as a comprehensive overview of the role of fuzzy logic in shaping the future of healthcare.

DOI: 10.4018/979-8-3373-0548-6.ch007

INTRODUCTION

Fuzzy Datasets and Soft Computing in Healthcare Informatics

Fuzzy datasets represent a critical paradigm in data science characterized by inherent imprecision, ambiguity, and ill-defined boundaries. Unlike traditional crisp datasets with clear-cut classifications, fuzzy datasets exhibit gradual, progressive membership and overlapping domains. Typical examples include sensor readings, textual data, human input data, and particularly, healthcare-related information (Zadeh, 1965). Fuzzy dataset is a data that is inherently imprecise or ambiguous, with ill-defined boundaries, overlapping domains, vague (non-crisp) sets, and gradually progressive membership. Sensor readings, textual dataset, human input data are few classical examples of Fuzzy datasets. Healthcare data like many biological data is inherently fuzzy in nature and such a data lends itself for analysis using Fuzzy logic (Kiseliova & Pagava, 2014).

Healthcare data exemplifies the quintessential fuzzy dataset, presenting complex, nuanced information that defies straightforward categorization. Biological and clinical data are inherently stochastic, with continuous variations that challenge conventional binary classification methods (Kiseliova & Pagava, 2014).

Conceptual Framework

Fuzzy logic provides a sophisticated mathematical framework for handling uncertainty and imprecision. Unlike traditional Boolean logic with strict true/false dichotomies, fuzzy logic allows for degrees of truth, enabling more nuanced data interpretation (Zadeh, 1994).

Applications in Healthcare

Fuzzy logic demonstrates exceptional utility in various healthcare management scenarios:

1. Clinical Diagnosis: Facilitating more sophisticated diagnostic processes by accommodating complex, non-linear medical data
2. Patient Intervention: Enabling personalized treatment strategies based on probabilistic patient profiles
3. Epidemiological Analysis: Supporting complex statistical modeling of disease progression and population health trends

Comparative Advantages

Compared to conventional mathematical programming techniques like linear programming, fuzzy logic offers:

- Enhanced flexibility in handling ambiguous data
- Reduced computational complexity
- More human-like reasoning capabilities
- Better representation of real-world medical uncertainties

Fuzzy logic approach can best handle the inherent 'fuzziness' in healthcare data. It can help in various healthcare management scenarios like clinical diagnosis or patient intervention. Fuzzy analysis can be employed in diagnostic, laboratory, and other epidemiological data (Kiraz, 2014). In contrast, using conventional mathematical programming techniques like Linear programming, may require considerable time, effort and other resources which maybe scarce in constrained settings. Typical example of Fuzzy techniques in clinical practice was discussed by Kiseliova and Pagava (2014) who used this approach in improving clinical decision-making in the specialty of Pediatrics (Child healthcare) (Kiseliova & Pagava, 2014).

Soft Computing: An Integrated Computational Approach

Soft computing is a discipline that deals with developing algorithms and creating models that have inherent abilities to deal with fuzzy data, which is imprecise or vague, and at times only partially accurate. This also includes various artificial intelligence frameworks like Genetic algorithms (GA), Artificial neural networks (ANNs), Rough sets, Evolution strategies, Bayesian net, Support vector machines (SVMs), and Fuzzy logic. Fuzzy logic approaches a principle component of the soft computing, has ability to handle of fuzzy data which are ubiquitous in the healthcare systems (Zadeh, 1965). Complex medical databases which store healthcare information like: Picture archiving communication system packs (PACS) or Electronic health records (EHRs) or Medical literature databases like Pubmed or Genomic databases like NCBI.This approach integrates multiple artificial intelligence paradigms namely:

- Genetic Algorithms (GA)
- Artificial Neural Networks (ANNs)
- Rough Sets
- Evolution Strategies
- Bayesian Networks

- Support Vector Machines (SVMs)
- Fuzzy Logic

An important application of Fuzzy logic in healthcare, is Fuzzy query to retrieve complex information from medical databases. Due to complex nature of such a database a conventional query would not yield useful results. Such a search information retrieval would be very useful in contrast to conventional query response paradigm in medical decision support system (Zadeh, 1965).

Information Retrieval Strategies

Fuzzy logic enables sophisticated information retrieval from complex medical databases through three primary approaches:

Rapid and efficient, vague information retrieval from medical databases can be achieved by Fuzzy sets. There are 3 approach that have been published in the literature for information retrieval:

1. Fuzzy Query Language Development

   This enables more flexible, context-aware search mechanisms and also accommodates semantic and contextual nuances in medical terminology
2. Fuzzy Relational Database Management Systems

   This system supports complex data relationships while handling inherent uncertainties in medical data storage and retrieval
3. Fuzzy Ontology-Based Information Modeling

   This is a system that creates sophisticated knowledge representation frameworks, and facilitates advanced semantic understanding of medical concepts (Zadeh, 1965).

Fuzzy Match Operator Methodologies

For example, the fuzzy match operator ~, permits matching words and phrases with similar sounding or similar syntactic meaning words in the medical databases. The fuzzy match operator (~) provides advanced matching capabilities across three dimensions.

1. String similarity like for example '*myo*' meaning muscle of heart or skeletal muscle
2. Phonetic similarity like for example '*myocardial*' with '*myocarditis*'
3. Semantic similarity like for example 'congestive cardiac failure' and 'cardiac insufficiency' (Zadeh, 1965).

The uncertainties and imprecisions that are inherently a part of healthcare data are best handled using a fuzzy approach. This approach can be used for handling various complex healthcare data sets like those produced from diagnostic, therapeutic, laboratory investigations, clinical medicine, patient records etc. Further, fuzzy queries are powerful tool for information retrieval from such complex medical databases. As a field of active research fuzzy approaches are likely to find more applications in the field of healthcare leading to improved clinical decision-making, and personalized patient care. Fuzzy logic and soft computing represent transformative approaches to managing healthcare's inherent complexity. By embracing uncertainty and developing sophisticated computational frameworks, researchers and clinicians can unlock deeper insights from increasingly complex medical datasets.

BACKGROUND

Foundations of Fuzzy Logic

Fuzzy logic Foundations of Fuzzy Logic

Fuzzy logic concept was introduced by Lotfi A. Zadeh in 1965. Unlike a traditional binary yes/ no, on / off, one / zero system fuzzy logic is a multivalued logic system which deals with ambiguous uncertain information. Zadeh (1965) states the definition of a fuzzy set as "the extension to classical sets of concepts which are not precisely definable but roughly or approximately described" (cited directly Zadeh, 1965, p. 338) (Zimmermann, 2001)

Fuzzy logic allows intermediate truth values between zero and one. This allows many real-world scenarios which have many intermediary states instead of a well-defined zero and one. This section further outlines the fundamentals of fuzzy logic (Abaei, R., & Kordrostami, S. 2022).

Basic Concepts

Fuzzy set is an extension of classic mathematical set for interpretation of imprecise inputs. Membership functions assign the degree of truth, by mapping the input value to membership degrees between zero and one (Buckley & Shensa, 1985).

Membership Functions

Membership functions are ones that define the boundary domains of the fuzzy sets. Membership functions have 4 parameters namely: Height, Boundary, Support, and Core. Support refers to range of input values, while core refers to central values. Boundary refers to transition zone, while Height represents the membership degree. The membership function supports 3 different set operators: union(max), intersection (min), and compliment (1-membership) (Buckley & Shensa, 1985).

Rule-Based Reasoning

A rule-based reasoning system uses a set of predefined rules to analyse the inputs and to arrive at definite conclusions. Fuzzy rules consist of proceeding antecedents and succeeding consequents (Klir & Yuan, 1995).

Antecedents are all the preconditions that must be satisfied and Consequents are conclusions or actions to be taken. Linguistic variables are set of Fuzzy variables with linguistic terms like 'best' or 'worst'. Defuzzification refers to the process of back converting fuzzy outputs to crisp values after processing. The fuzzy RBR Process consists of Fuzzification, Rule evaluation, Inference, and Defuzzification (Chen & Gupta, 1998).

Fuzzy Logic Implementations

There are numerous applications of fuzzy logic in various disciplines including engineering science, market research, financial management, medical science, and automation. Applied examples include:

- Expert systems (Hwang & Sivanandam, 2002)
- Intelligent control systems (Mamdani, 1974)
- Decision support systems (Mamdani, 1974)
- Signal processing (Pal & Mitra, 2005)
- Image processing (Ross, 2010)
- Natural language processing (Ross, 2010)
- Expert systems (Jang et al., 1993)

The computational complexity, scalability and interpretability have been disconcerting about the widespread adoption of fuzzy logic (Klir & St. Clair, 1997). To circumvent this hurdle, research has focused on adoption of advanced optimization algorithms, deployment of hybrid systems and the use of artificial intelligence and machine learning techniques (Yager & Filev, 1994).

The Need for Fuzzy Approaches in Healthcare

Fuzzy logic has been increasingly utilized in the healthcare sector due to its ability to handle uncertainty and imprecision inherent in medical practice. This chapter aims to provide a comprehensive overview of current applications of fuzzy logic in medicine and bioinformatics, focusing on medical literature reviews will be discussed, along with few examples of applications in medicine.

Diagnosis in medicine often involves multiple levels of uncertainty and imprecision, making fuzzy logic an appropriate approach to address these issues (Chung et al., 2013). Fuzzy logic can be considered a qualitative computational method that introduces partial truth, allowing for a more nuanced representation of reality. In healthcare, fuzzy set theory can be used to handle the intrinsic imprecision of many biomedical and bioinformatics problems.

Fuzzy logic has been applied to various disease groups, including predicting treatment responses in alcohol dependence, dosage determination, and studying fuzzy epidemics. A search using the MEDLINE database identified a total of 804 articles between 1991 and 2002 that employed fuzzy logic in medical contexts. The number of articles has shown exponential growth over the years, highlighting the increasing significance of fuzzy logic in healthcare research (De Morgan, 1860).

Fuzzy logic offers a valuable approach to handling uncertainty and imprecision in healthcare, making it a suitable tool for various medical applications. As medical data continues to grow in complexity, the role of fuzzy logic in healthcare is likely to expand further.

Fuzzy logic is a mathematical approach that deals with imprecision and uncertainty, allowing for the representation of vague or ambiguous concepts using fuzzy sets and fuzzy rules. Unlike classical binary logic, where a statement is either true or false, fuzzy logic recognizes that many real-world situations involve degrees of truth or membership. This concept has found various applications in various domains, including healthcare. The following section presents the fundamentals of fuzzy logic and illustrates its applications in healthcare with relevant examples (De Morgan, 1860).

Fuzzy Sets and Membership Functions

The core concept of fuzzy logic is the fuzzy set, which is a generalization of the traditional crisp set. In a crisp set, an element either belongs to the set or does not belong to it. However, in a fuzzy set, an element can have a partial membership, represented by a membership function that assigns a value between 0 and 1 to each element (Kangasniemi et al., 2017). The membership function quantifies the degree to which an element belongs to the fuzzy set.

Fuzzy Rules and Fuzzy Inference Systems

Fuzzy rules are IF-THEN statements that relate fuzzy inputs to fuzzy outputs. These rules are based on expert knowledge or empirical data and are used to reason about imprecise or uncertain situations. A fuzzy inference system (FIS) combines these fuzzy rules with fuzzy sets to produce an output based on the given inputs (Kangasniemi et al., 2017).

Applications in Healthcare

1. **Medical Diagnosis and Decision Support Systems**: Fuzzy logic has been applied in medical diagnosis and decision support systems to handle the inherent uncertainty and vagueness in medical data and knowledge. For example, a fuzzy system can assist in the diagnosis of heart disease by considering various risk factors, such as age, blood pressure, and cholesterol levels, which may have imprecise or overlapping boundaries (Peirce, 1887).
2. **Patient Monitoring and Alert Systems**: Fuzzy logic can be used in patient monitoring systems to interpret and analyze vital signs and other physiological data. By defining fuzzy rules based on expert knowledge, these systems can generate alerts or recommendations for medical intervention. For instance, a fuzzy system could analyze the combination of heart rate, blood pressure, and respiratory rate to detect potential complications or deterioration (Farhadian et al., 2018).
3. **Medical Image Processing and Analysis**: Fuzzy logic techniques have been employed in medical image processing and analysis tasks, such as image segmentation, enhancement, and feature extraction. Fuzzy approaches can handle the inherent ambiguity and uncertainty in medical images, resulting in more accurate and robust analysis. For example, fuzzy clustering algorithms can be used for tumor segmentation in brain MRI images (Liao et al., 2019).
4. **Medical Decision-Making and Risk Assessment**: Fuzzy logic can aid in medical decision-making and risk assessment by considering multiple factors and their degrees of importance or uncertainty. For instance, a fuzzy system could evaluate the risk of developing a particular disease based on various risk factors, such as family history, lifestyle, and environmental exposures (John et al., 2014).

Rule-based reasoning in fuzzy logic

Rule-based reasoning in fuzzy logic is a powerful tool for handling uncertainty and imprecision in knowledge representation and decision-making. It combines the flexibility of fuzzy logic with the structured approach of rule-based systems, allowing for more human-like reasoning in complex systems (Turky & Hassanien, 2014).

Fuzzy Rules and Fuzzy Inference Systems in Healthcare

Fuzzy rules and fuzzy inference systems are essential components of fuzzy logic that enable reasoning and decision-making in the presence of uncertainty and imprecision. These concepts have found numerous applications in the healthcare domain, where ambiguity and vagueness are often encountered.

Fuzzy Rules

Fuzzy rules are IF-THEN statements that relate fuzzy inputs to fuzzy outputs. These rules are constructed based on expert knowledge or empirical data and are used to represent the relationship between different variables or factors in a given problem domain. In healthcare, fuzzy rules can capture the complexity and nuances of medical knowledge and decision-making processes.

Example: Consider a fuzzy rule for assessing the risk of developing gestational diabetes mellitus (GDM) during pregnancy (Kiseliova & Pagava, 2014):

IF (maternal age is high) AND (pre-pregnancy BMI is overweight) AND (family history of diabetes is positive) THEN (risk of GDM is high)

In this rule, the antecedents (maternal age, pre-pregnancy BMI, and family history of diabetes) are fuzzy inputs, and the consequent (risk of GDM) is the fuzzy output. The linguistic terms "high," "overweight," "positive," and "high risk" are defined by corresponding membership functions, allowing for partial membership and gradual transitions.

Fuzzy Inference Systems (FIS)

A fuzzy inference system (FIS) is a computational framework that combines fuzzy rules and fuzzy sets to produce an output based on given inputs. The FIS follows a set of steps, including fuzzification, rule evaluation, and defuzzification, to make decisions or generate outputs based on the provided fuzzy inputs.

Example: A fuzzy inference system can be used for monitoring and adjusting insulin delivery in patients with type 1 diabetes mellitus. The inputs to the system may include blood glucose levels, food intake, and physical activity, while the output is the recommended insulin dose (Kiraz, 2014).

The steps involved in this FIS can be as follows:

1. **Fuzzification:** Convert the crisp input values (e.g., blood glucose level of 180 mg/dL) into fuzzy sets using membership functions.
2. **Rule evaluation:** Apply the fuzzy rules (e.g., IF blood glucose is high AND physical activity is low THEN insulin dose is high) to the fuzzy inputs and determine the degree of truth for each rule.
3. **Aggregation:** Combine the outputs of all the rules using fuzzy set operations (e.g., union or intersection).
4. **Defuzzification:** Convert the aggregated fuzzy output back into a crisp value (e.g., 12 units of insulin) using a defuzzification method (e.g., centroid or mean of maximum).

The FIS can continuously monitor the inputs and adjust the insulin dose accordingly, providing a more personalized and adaptive approach to diabetes management.

Rule-Based Reasoning

Rule-based reasoning is a problem-solving approach where knowledge is represented in the form of IF-THEN rules. These rules capture relationships between variables and provide a structured way to reason about a problem domain. In rule-based systems, the knowledge base consists of a set of rules, and the reasoning process involves applying these rules to a given set of facts to reach a conclusion (Yager & Filev, 1993).

When fuzzy logic is incorporated into rule-based reasoning, it becomes a powerful framework for handling complex, uncertain situations. Fuzzy rules allow for imprecise or vague conditions and conclusions, providing a more flexible representation of knowledge. For example, consider a simple rule in a traditional rule-based system:

- IF temperature is high, THEN turn on the air conditioner.

- In a fuzzy rule-based system, this rule can be expressed in a more nuanced manner:
- IF temperature is somewhat high, THEN turn on the air conditioner at a low speed.
- IF temperature is very high, THEN turn on the air conditioner at a high speed.

Here, the terms "somewhat high" and "very high" are fuzzy sets that can have varying degrees of membership, depending on the actual temperature value. The speed of the air conditioner can also be a fuzzy variable, allowing for a smooth transition between different speeds (Li et al., 2019).

In the examples provided, fuzzy logic has been applied to various healthcare domains, such as medical diagnosis, patient monitoring, image analysis, and decision-making. The ability of fuzzy logic to handle imprecision and uncertainty makes it a valuable tool for addressing the complexities and ambiguities inherent in healthcare scenarios.

Advantages of Fuzzy Rule-Based Systems

1. Flexibility: Fuzzy rule-based systems can handle imprecise and uncertain information, making them suitable for domains where precise measurements or definitions are difficult to obtain.
2. Human-like reasoning: Fuzzy rules can mimic human reasoning patterns, making the system more interpretable and acceptable to users.
3. Knowledge representation: Fuzzy rules provide a structured way to represent and organize knowledge, making it easier to maintain and update the system.
4. Smooth decision-making: Fuzzy logic allows for gradual transitions between states, avoiding abrupt changes that may be unrealistic in certain applications (Kangasniemi et al., 2017).

Applications of Fuzzy Rule-Based Systems

1. Fuzzy rule-based systems have been applied in various domains, including:
2. Control systems: Fuzzy logic controllers are used in industrial processes, robotics, and autonomous vehicles to handle complex, nonlinear behaviours.
3. Image processing: Fuzzy rules can be used for image segmentation, edge detection, and pattern recognition tasks.
4. Natural language processing: Fuzzy logic can be applied to handle ambiguity and uncertainty in language understanding and machine translation.
5. Medical diagnosis: Fuzzy rule-based systems can assist in medical decision-making by incorporating imprecise symptoms and patient characteristics.

6. Financial forecasting: Fuzzy logic can be used to model complex relationships and uncertainties in financial markets (Peirce, 1887), (Farhadian et al., 2018).

Rule-based reasoning in fuzzy logic provides a powerful framework for handling uncertainty and imprecision in knowledge representation and decision-making. By combining the strengths of fuzzy logic and rule-based systems, this approach enables more flexible and human-like reasoning in a wide range of applications.

Application of Fuzzy Logic in Medical Diagnosis

The application of fuzzy logic in medical diagnosis offers a valuable approach to handling the inherent uncertainty and imprecision often present in clinical decision-making. By incorporating fuzzy set theory, medical diagnostic systems can better reflect the complex and vague nature of medical knowledge, leading to improved accuracy and reliability (Liao et al., 2019).

Medical diagnosis refers to the process of identifying a disease or condition based on a patient's symptoms, medical history, laboratory tests, and other relevant data. It is a complex task that requires knowledge, experience, and the ability to interpret often ambiguous information. Traditional diagnostic methods rely on crisp thresholds and categorical decisions, which may not adequately capture the nuances of certain medical conditions (John et al., 2014).

Challenges in Medical Diagnosis

1. Imprecision: Medical data can be imprecise and vague. Symptoms may vary widely among patients, and the boundaries between normal and abnormal conditions may not always be clear-cut.
2. Incomplete Information: In many cases, not all relevant information is available, or it may be difficult to obtain. Missing or incomplete data can make diagnosis challenging.
3. Uncertainty: Medical knowledge is often probabilistic and uncertain. The presence or absence of certain symptoms may indicate a particular disease with varying degrees of likelihood.
4. Individual Variability: Different patients may exhibit varying responses to the same disease, making it difficult to apply a one-size-fits-all approach to diagnosis (John et al., 2014).

Fuzzy Logic in Medical Diagnosis

Fuzzy logic provides a framework to model and handle the uncertainty and imprecision inherent in medical diagnosis. It allows for gradual transitions between categories and the representation of vague concepts. Here's how fuzzy logic can be applied in medical diagnostic systems:

1. Fuzzy Sets: Fuzzy sets can be used to represent symptoms, laboratory test results, and other medical parameters. Instead of crisp categories (e.g., normal or abnormal), fuzzy sets allow for degrees of membership. For example, a patient's temperature can be described as slightly elevated, moderately elevated, or highly elevated, rather than simply normal or abnormal.
2. Fuzzy Rules: Fuzzy rules can capture the relationships between symptoms, risk factors, and diseases. These rules can be derived from medical knowledge, expert opinions, or data-driven approaches. For instance, a rule might state that IF the patient has a severe headache AND a fever, THEN the likelihood of influenza is high.
3. Defuzzification: In the decision-making process, fuzzy logic can provide a continuous output, such as the likelihood of a particular disease. Defuzzification techniques can then be applied to obtain a crisp diagnosis or a recommended course of action (Turky & Hassanien, 2014).

Advantages of Fuzzy Logic in Medical Diagnosis

1. Improved Accuracy: Fuzzy logic can better represent the complex relationships between symptoms and diseases, leading to more accurate diagnoses, especially in cases where traditional methods struggle with ambiguity.
2. Handling Incomplete Data: Fuzzy logic can provide reasonable conclusions even with incomplete or uncertain data, helping clinicians make informed decisions in the absence of all relevant information.
3. Incorporating Expert Knowledge: Fuzzy rule-based systems can encode the knowledge and experience of medical experts, making the diagnostic process more consistent and less prone to individual biases.
4. Interpretable Results: Fuzzy rules provide a transparent representation of the reasoning process, enhancing trust and understanding of the diagnostic system's output (Yager & Filev, 1993).

Applications in Medical Diagnosis

1. Clinical Decision Support: Fuzzy logic can be integrated into clinical decision support systems to assist physicians in diagnosing diseases, such as diabetes, cardiovascular disorders, or infectious diseases.
2. Differential Diagnosis: Fuzzy logic can aid in differential diagnosis by evaluating the likelihood of multiple possible diseases based on a patient's symptoms.
3. Disease Risk Assessment: Fuzzy models can be used to assess an individual's risk of developing certain diseases, considering genetic, environmental, and lifestyle factors.
4. Personalized Medicine: Fuzzy logic can facilitate personalized treatment plans by considering individual variability and responses (Li et al., 2019).

The application of fuzzy logic in medical diagnosis offers a promising approach to addressing the challenges of uncertainty and imprecision. By incorporating fuzzy set theory, medical diagnostic systems can better reflect the complexities of clinical decision-making, leading to improved patient care and outcomes. Further research and development in this area have the potential to revolutionize the practice of medicine (Li et al., 2019).

The examples illustrate how fuzzy rules and fuzzy inference systems can capture the complexity and uncertainty inherent in healthcare scenarios, such as assessing the risk of gestational diabetes or adjusting insulin doses for diabetes management. By incorporating expert knowledge and handling imprecision, these fuzzy logic techniques can provide valuable decision support and adaptive control in healthcare applications (Chen et al., 2014).

Graphical illustrations of Fuzzy approaches in healthcare

Fuzzy approaches in healthcare can be illustrated using various graphs and visual representations. Here are some examples of graphs along with their key points:

Membership Functions: These are Graphs representing membership functions are fundamental in fuzzy logic systems. They visually depict the degree of membership (y-axis) for different values or ranges of input variables (x-axis). Common membership function shapes include triangular, trapezoidal, Gaussian, and bell-shaped curves. Key points are represented for vague or imprecise concepts, allow for gradual transitions and partial memberships (Prasad et al., 2010).

Figure 1. Illustration of various membership functions, these can be, triangular, trapezoidal, Gaussian, and bell-shaped curves, original compilation of the authors in seaborn module

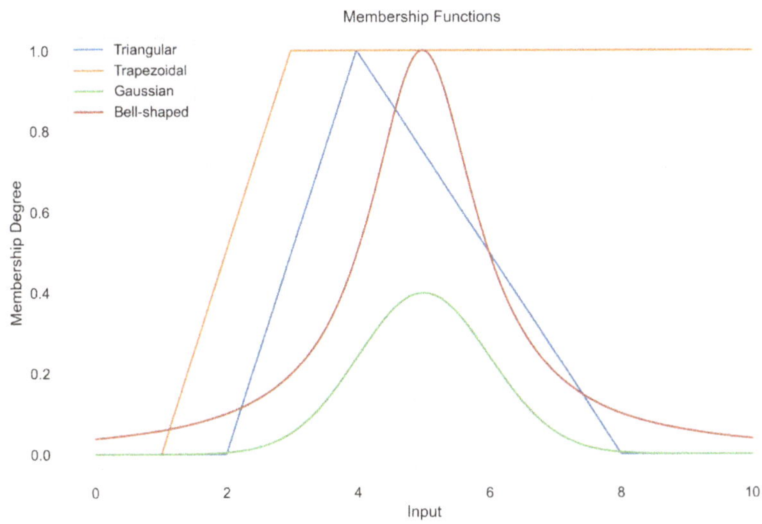

Rule Surfaces: Fuzzy rule surfaces (or control surfaces) graphically represent the behavior of a fuzzy inference system with two inputs and one output. They provide a 3D visualization of the relationship between input variables and the output variable based on the defined fuzzy rules. These surfaces can help in understanding and analyzing the overall system behavior. Key points are visualization of the impact of input variable combinations on the output, aid in rule interpretation and analysis (Lilly, 2010).

Figure 2. Illustration of Fuzzy rule surfaces (or control surfaces) which outlines the performance of a fuzzy inference system with two inputs and one output. In this example, 3D graph the severity of a cardiovascular disease is based on 2 sample variable namely age and blood pressure. Original illustration of the authors using matplotlib module.

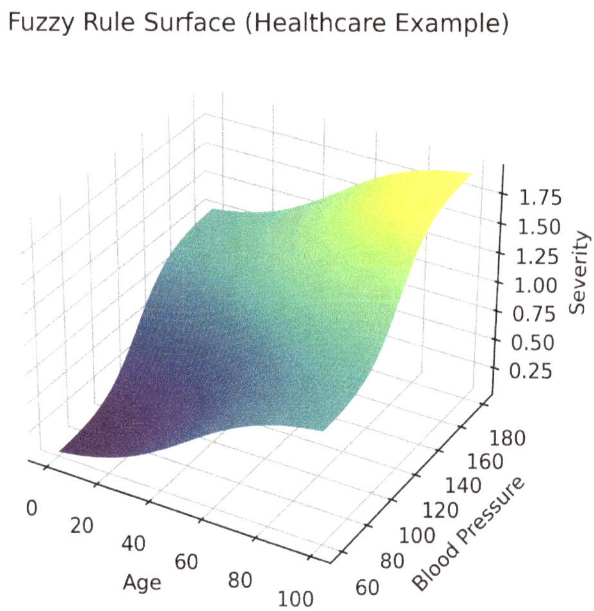

Fuzzy Clustering: These are Graphs can be used to visualize the results of fuzzy clustering algorithms applied to medical data or images. Scatter plots can represent data points with different colors or symbols indicating their cluster assignments. Line graphs or heat maps can illustrate the membership degrees of data points to different clusters. Key points are visualization of data patterns and cluster structures, aid in interpreting clustering results (Pham & Prince, 1999).

Fuzzy Risk Assessment: These are Graphs can be used to represent the risk assessment or risk stratification results obtained from fuzzy inference systems. Bar charts or line graphs can illustrate the relative risk levels or probabilities for different outcomes or conditions. Multiple bars or lines can represent the risk levels for different patient subgroups or risk factor combinations. Key points are visualization of risk levels, compare risk profiles, aid in risk communication and decision-making (Naaz et al., 2011).

Figure 3. Bar chart, illustration of Fuzzy Risk Assessment of cardiovascular disease based on multiple factors like High blood pressure, serum cholesterol levels, presence of type 2 diabetes mellitus, obesity, and a positive family history. Original graph of the authors using seaborn module and data from the literature mostly systematic reviews.

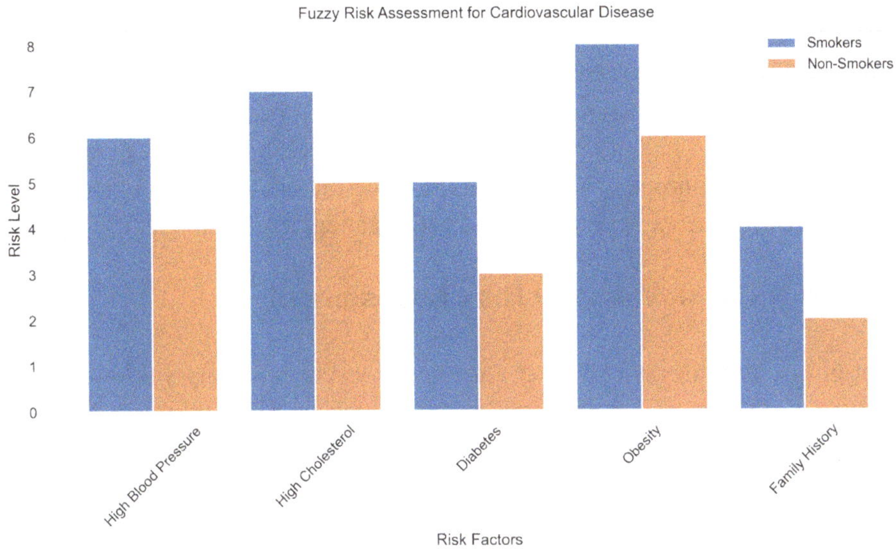

These graphs serve as visual aids in understanding and interpreting the results of fuzzy approaches in healthcare. They can facilitate communication among healthcare professionals, support decision-making processes, and provide insights into the behavior and performance of fuzzy logic systems in various healthcare applications.

PRACTICAL APPLICATIONS OF FUZZY LOGIC IN MEDICAL DIAGNOSIS

The application of fuzzy logic in medical diagnosis offers a valuable approach to handling uncertainty and imprecision inherent in clinical decision-making. By incorporating fuzzy set theory and fuzzy rule-based systems, medical experts can model complex relationships and provide more nuanced assessments, leading to improved diagnostic accuracy and patient care. Here are some examples of how fuzzy logic has been applied in medical diagnosis:

Example 1: Diabetes Diagnosis

Diabetes mellitus is a complex metabolic disorder that requires careful diagnosis and management. Traditional diagnostic criteria rely on fasting plasma glucose or oral glucose tolerance tests with rigid cut-off points. However, these cut-off points may not capture the variability in glucose metabolism among individuals. Fuzzy logic can be used to develop a more flexible diagnostic system. Yadav et al. proposed a fuzzy expert system for the diagnosis of diabetes mellitus (Naaz et al., 2011). They defined fuzzy rules based on multiple factors, including fasting plasma glucose, postprandial glucose, and glycated haemoglobin (HbA1c) levels. The system provides a continuous diabetes index value that indicates the likelihood of diabetes, allowing for early detection and better patient management.

Example 2: Cardiovascular Risk Assessment

Cardiovascular disease (CVD) is a leading cause of mortality worldwide, and accurate risk assessment is crucial for preventive measures. Traditional risk assessment models, such as the Framingham Risk Score, use discrete cut-off points for risk factors, which may not fully capture individual variations. Fuzzy logic can help in refining these risk assessments.

Li et al. developed a fuzzy rule-based system for CVD risk assessment (Khammas et al., 2020). Their system incorporated fuzzy rules for multiple risk factors, including age, gender, blood pressure, cholesterol levels, smoking status, and family history. By considering the interactions and uncertainties among these factors, the system provided a more personalized risk estimate, aiding in early intervention and targeted treatment (Gupta, R., et. al., 2023).

Example 3: Breast Cancer Diagnosis

Breast cancer diagnosis often involves evaluating various factors, such as mammography findings, ultrasound results, clinical symptoms, and patient history. Integrating these factors to reach a diagnosis can be challenging due to their subjective nature and potential overlaps between malignant and benign lesions. Fuzzy logic can assist in combining these factors for a more comprehensive assessment.

Sohrabi et al. proposed a fuzzy expert system for breast cancer diagnosis (Mausoor & Graham, 2007). Their system incorporated fuzzy rules based on clinical symptoms, imaging findings, and biopsy results. By assigning different weights to each factor and considering their interactions, the system provided a more accurate diagnosis, helping clinicians make informed treatment decisions.

Example 4: Neurological Disorder Diagnosis

Neurological disorders, such as Alzheimer's disease and Parkinson's disease, present with a range of symptoms and can be challenging to diagnose accurately. Fuzzy logic can help in integrating clinical symptoms, laboratory tests, and imaging findings to support differential diagnosis and disease staging.

Chakraborty et al. developed a fuzzy expert system for the diagnosis of Alzheimer's disease (Mamdani & Assilian, 1975). Their system considered clinical symptoms, cognitive test scores, and cerebrospinal fluid biomarkers to provide a probabilistic diagnosis. This approach aids in early detection, differential diagnosis, and monitoring of disease progression.

These examples illustrate the potential of fuzzy logic in medical diagnosis, where uncertainty and imprecision are prevalent. By incorporating fuzzy set theory and rule-based systems, clinicians can make more informed decisions, provide early interventions, and improve patient outcomes. Further research and development in this field hold promise for advancing personalized medicine and enhancing the accuracy and effectiveness of medical diagnosis.

Advantages of Fuzzy Logic in Medical Diagnosis

Fuzzy logic has gained significant attention in the field of medical diagnosis due to its ability to handle uncertainty and imprecision, providing several advantages over traditional binary decision-making approaches. Here is an overview of the advantages of fuzzy logic in medical diagnosis:

Handling Uncertainty: Medical diagnosis often involves uncertain and imprecise information. Fuzzy logic allows for the representation of vague concepts and symptoms, providing a more realistic model of medical knowledge. This is particularly useful when dealing with symptoms that are difficult to quantify precisely, such as "mild", "moderate", or "severe" pain (Mendel, 1995).

Flexibility in Knowledge Representation: Fuzzy logic provides a flexible framework for representing medical knowledge. It can incorporate linguistic variables and rules that mirror the way physicians think and reason. This facilitates the translation of qualitative medical knowledge into quantitative models, making the system more interpretable and acceptable to medical practitioners (Wang et al., 2016).

Gradual Decision-Making: Traditional binary decision-making often results in abrupt transitions between states, which may not reflect the gradual nature of certain medical conditions. Fuzzy logic allows for smooth transitions between categories, providing a more nuanced representation of disease progression. This can lead to more accurate diagnoses and treatment plans (Liu et al., 2019).

Incorporating Expert Knowledge: Fuzzy logic systems can effectively capture and utilize the knowledge and experience of medical experts. By representing expert knowledge in the form of fuzzy rules, the system can emulate the decision-making process of experienced physicians. This knowledge-based approach can be particularly valuable in rare or complex cases where data may be scarce (Chou et al., 2013).

Integration with Other Techniques: Fuzzy logic can be seamlessly integrated with other artificial intelligence and machine learning techniques. For example, it can be combined with neural networks, decision trees, or genetic algorithms to enhance their ability to handle uncertainty. This hybrid approach can lead to more robust and accurate diagnostic systems (Weber & Mandl, 2022).

Patient-Specific Diagnosis: Fuzzy logic can tailor diagnoses and treatment plans to individual patients. By considering multiple factors, including symptoms, medical history, and laboratory results, fuzzy logic systems can provide personalized assessments. This patient-centric approach can improve the accuracy of diagnoses and enhance patient care. (Liu et al., 2021)

Explanation and Transparency: Fuzzy logic systems can provide explanations for their diagnoses, enhancing transparency and trust. The use of linguistic variables and rules makes it easier to understand the reasoning behind a particular conclusion. This is especially important in the medical field, where explainability and accountability are crucial (Chen et al., 2022).

Adaptability and Learning: Fuzzy logic systems can adapt to new knowledge and changing conditions. They can incorporate feedback and learn from experience, refining their rules and improving performance over time. This adaptability is essential in the dynamic field of medicine, where new research and treatments are constantly emerging (Wang et al., 2018).

Fuzzy logic offers significant advantages in medical diagnosis by providing a flexible and nuanced approach to handling uncertainty and imprecision. Its ability to emulate human reasoning, incorporate expert knowledge, and adapt to new information makes it a valuable tool in decision support systems for healthcare professionals. Further research and development in this area have the potential to improve diagnostic accuracy, enhance patient care, and contribute to better health outcomes.

LIMITATIONS OF FUZZY LOGIC IN MEDICAL DIAGNOSIS

The use of fuzzy logic in medical diagnosis has gained attention due to its ability to handle uncertainty and imprecision inherent in medical data and knowledge. While fuzzy logic offers several advantages, there are also limitations and challenges that need to be addressed for its effective application in this critical domain. This section discusses the limitations of fuzzy logic in medical diagnosis and provides insights into potential solutions and future research directions.

1. Knowledge Acquisition and Representation

One of the primary challenges in fuzzy logic applications is the acquisition and representation of knowledge in a structured and meaningful way. In medical diagnosis, this challenge is amplified due to the complexity and dynamic nature of medical knowledge (Li et al., 2020). Acquiring knowledge from medical experts and representing it in the form of fuzzy rules or sets can be a time-consuming and error-prone process.

2. Subjectivity and Interpretability

Fuzzy logic relies on the definition of fuzzy sets and membership functions, which involve a degree of subjectivity (Liu et al., 2021). Different experts may define fuzzy sets and membership functions differently, leading to variations in the system's output. Interpretability is also a concern, as the reasoning process in fuzzy logic systems can be difficult to explain, especially to non-experts (Lee et al., 2019).

3. Handling High-Dimensional Data

Medical diagnosis often involves dealing with high-dimensional data, such as patient records with numerous variables. Fuzzy logic systems, particularly those based on rule-based reasoning, can become unwieldy as the number of variables increases (Coiera, 2022). The curse of dimensionality poses challenges in both knowledge representation and computational complexity.

4. Lack of Standardization

There is a lack of standardized methodologies and guidelines for developing fuzzy logic systems in medical diagnosis. This can lead to inconsistencies in system design and performance, making it difficult to compare and evaluate different approaches (Dubois & Prade, 1980).

5. Data Quality and Availability

The performance of fuzzy logic systems heavily depends on the quality and quantity of data used for training and validation. In medical diagnosis, obtaining large, labelled datasets can be challenging due to privacy concerns, ethical issues, and the cost of data collection (Zadeh, 1993). Additionally, missing data and outliers can affect the accuracy and reliability of fuzzy logic models.

6. Calibration and Tuning

Fuzzy logic systems require careful calibration and tuning of membership functions and rule weights to achieve optimal performance. This process can be time-consuming and may require extensive domain knowledge and experimentation (Yen & Lang, 2205). Overly complex membership functions or an excessive number of rules can lead to overfitting, while overly simple functions may result in underfitting.

Addressing the Limitations of Fuzzy Systems

For addressing the limitations of Fuzzy Systems, in healthcare scenarios, the following paradigms can be deployed as solution:

1. Hybrid Models

Combining fuzzy logic with other artificial intelligence techniques, such as machine learning, can help address some of the limitations. For example, machine learning algorithms can be used to automatically extract rules from data, reducing the burden of knowledge acquisition (Mamdani, 1974). Hybrid models can also improve the handling of high-dimensional data and enhance the performance of fuzzy logic systems (Rajamani & Iyer, 2024).

2. Standardization and Guidelines

Efforts should be made to develop standardized methodologies, frameworks, and evaluation metrics for fuzzy logic applications in medical diagnosis. This includes guidelines for knowledge acquisition, system design, validation, and reporting of results (John et al., 2011). Standardization will facilitate reproducibility, comparison of different approaches, and the establishment of best practices.

3. Explainable Fuzzy Systems

Research should focus on developing fuzzy logic systems that are more interpretable and explainable. This includes the use of transparent membership functions, rule visualization techniques, and post-hoc explanation methods (Bellman, 1961). Explainable fuzzy systems can enhance trust and acceptance among medical professionals and patients.

4. Data-driven Approaches

Advances in data collection and sharing, such as electronic health records and medical imaging databases, should be leveraged to address data-related limitations. Data-driven approaches, including deep learning with fuzzy logic, can be explored to handle high-dimensional data and improve the accuracy and robustness of fuzzy logic systems (Wong & Liu, 2008).

5. Collaborative Efforts

Collaborative efforts involving medical experts, data scientists, and fuzzy logic researchers are crucial for overcoming the limitations. Multidisciplinary teams can work together to address knowledge acquisition, system design, and validation challenges, ensuring that fuzzy logic systems are clinically relevant and effective.

Fuzzy logic has the potential to revolutionize medical diagnosis by providing a flexible framework for handling uncertainty. However, several limitations, including knowledge acquisition, subjectivity, high-dimensional data, and lack of standardization, need to be addressed. By exploring hybrid models, standardization, explainability, data-driven approaches, and collaborative efforts, the limitations can be mitigated, leading to more accurate, reliable, and widely accepted fuzzy logic applications in medical diagnosis (Wong & Liu, 2008).

Fuzzy logic has shown great potential in medical diagnosis by providing a flexible framework to handle uncertainty and vague information inherent in clinical decision-making. The ability of fuzzy logic to model complex relationships and interpret imprecise data makes it a valuable tool in this domain. Future research directions in the application of fuzzy logic to medical diagnosis aim to enhance its capabilities and broaden its impact. This article explores some of these promising avenues (Klir & Yuan, 1995).

FUTURE DIRECTIONS

Prospectively the following future directions are envisaged by the authors, in the application of Fuzzy logic in healthcare settings:

1. **Integration with Machine Learning**: Fuzzy logic can be combined with machine learning techniques to enhance diagnostic accuracy and adaptability. Hybrid models that incorporate fuzzy logic with artificial neural networks, decision trees, or support vector machines can leverage the strengths of both approaches. Machine learning algorithms can learn complex patterns from data,

while fuzzy logic can provide interpretability and handle uncertainty effectively. This integration has the potential to improve diagnosis accuracy, especially in dynamic and nonlinear medical domains (Tsoukias & Narayanan, 1994).

2. **Personalized Medicine:** Fuzzy logic can play a crucial role in realizing the vision of personalized medicine by tailoring diagnoses and treatments to individual patients. By incorporating patient-specific factors such as genetic information, lifestyle, and environmental exposures, fuzzy models can provide more precise and personalized assessments. This direction holds promise in fields like oncology, where treatment plans can be optimized based on an individual's unique characteristics (Li & Pedrycz, 2009).
3. **Multimodal Data Fusion:** Medical diagnosis often involves multiple sources of information, including clinical symptoms, laboratory tests, imaging data, and genetic profiles. Fuzzy logic can serve as a unifying framework for integrating and interpreting these diverse data modalities. By fusing information from multiple sources, fuzzy systems can provide more comprehensive and accurate diagnoses, reducing the potential for errors arising from single-modality assessments (Rajamani & Iyer, 2023).
4. **Explainable AI in Medicine:** Fuzzy logic contributes to the development of explainable artificial intelligence (XAI) in medicine. The interpretable nature of fuzzy rules enhances transparency and trust in diagnostic decisions made by AI systems. Future research can focus on designing fuzzy rule-based systems that provide justifiable explanations for their diagnoses, addressing ethical concerns and facilitating clinician acceptance (Jensen et al., 2012).
5. **Real-time Decision Support**: Fuzzy logic can be integrated into real-time decision support systems to assist clinicians during the diagnostic process. By providing timely and adaptive recommendations, these systems can improve the efficiency and accuracy of diagnoses, especially in time-critical situations. This direction holds promise for emergency medicine, intensive care, and remote healthcare settings (Mendel, 1995).
6. **Fuzzy Logic in Telemedicine:** With the growing adoption of telemedicine, fuzzy logic can enhance remote diagnostics and consultations. Fuzzy systems can assist in interpreting patient data transmitted remotely, providing initial assessments and recommendations. This can improve access to healthcare, especially in underserved or remote areas, and reduce the need for physical visits in certain cases (Lin et al., 2006).
7. **Ethical and Legal Considerations**: As fuzzy logic becomes more prevalent in medical diagnosis; ethical and legal implications arise. Future research should address issues such as accountability, privacy, and fairness in fuzzy logic-based diagnostic systems. Guidelines and regulatory frameworks need to be developed

to ensure the responsible use of this technology in healthcare (John & Shankar, 2012).

8. **Standardization and Interoperability:** Standardization of fuzzy logic-based diagnostic systems is crucial for their widespread adoption and integration into existing healthcare infrastructure. Future efforts should focus on developing standardized frameworks, ontologies, and data formats to ensure interoperability between different systems and facilitate data exchange (Ribeiro et al., 2016).

CONCLUSION: FUZZY LOGIC AS A PARADIGM SHIFT IN MEDICAL INFORMATICS

The trajectory of fuzzy logic in medical diagnosis represents far more than a mere technological advancement; it embodies a fundamental reimagining of how healthcare systems process, interpret, and leverage complex medical information. As healthcare continues to evolve in an era of unprecedented technological convergence, fuzzy logic emerges as a critical bridge between computational complexity and clinical intuition, offering a sophisticated framework for navigating the inherent uncertainties of medical decision-making (Estrada-Reyes et. al., 2021).

The potential for transformative impact extends across multiple dimensions of healthcare delivery. By providing flexible, adaptive decision-making tools, fuzzy logic addresses a critical gap in traditional diagnostic approaches that often struggle with the nuanced, non-linear nature of human biological systems. Unlike binary computational models, fuzzy logic accommodates the gradational, contextual nature of medical data, mirroring the complex reasoning processes of experienced clinicians (Das S., et. al., 2022). The explainability of fuzzy logic models emerges as a paramount concern. As healthcare becomes increasingly data-driven, there is a growing imperative to develop transparent, interpretable computational approaches that clinicians can trust.

Critical research directions demand a multifaceted approach to further developing fuzzy logic applications. Integration challenges require sophisticated computational frameworks that can seamlessly interface with existing healthcare information systems. Personalization strategies must evolve to leverage individual patient data more comprehensively, creating truly precision medicine approaches. Multimodal data fusion represents a particularly promising frontier, enabling comprehensive analysis across diverse data types—from genomic sequences to clinical imaging and patient-reported outcomes (Chen, Y., Liu, X., et. al., 2023).

The explainability of fuzzy logic models emerges as a paramount concern. As healthcare becomes increasingly data-driven, there is a growing imperative to develop transparent, interpretable computational approaches that clinicians can trust and understand. This necessitates innovative methodological developments that balance computational sophistication with clinical intuition, creating decision support tools that are both powerful and comprehensible.

Real-time decision support capabilities present another critical area of potential transformation. The ability to process complex, evolving medical data instantaneously could revolutionize clinical interventions, enabling more proactive, preventative approaches to patient care. Telemedicine, accelerated by recent global health challenges, represents an ideal context for deploying advanced fuzzy logic methodologies, potentially democratizing access to sophisticated medical diagnostics.

Ethical considerations and standardization remain crucial challenges that require ongoing interdisciplinary collaboration. Developing robust governance frameworks, ensuring patient privacy, and creating standardized protocols for fuzzy logic implementation will be essential to widespread adoption. This demands a collaborative approach involving clinicians, data scientists, ethicists, and regulatory bodies.

The ultimate promise of fuzzy logic in healthcare transcends technological innovation. It represents a profound shift towards more nuanced, personalized, and adaptive medical practices. By embracing computational approaches that can handle uncertainty, capture contextual complexity, and support more individualized patient care, we move closer to a healthcare paradigm that is simultaneously more precise and more human-entered.

Looking forward, the research trajectory suggests exponential potential. Emerging technologies like quantum computing, advanced machine learning, and sophisticated artificial intelligence frameworks are likely to further enhance fuzzy logic's capabilities. The convergence of these technologies promises not just incremental improvements, but potentially revolutionary changes in how we understand, predict, and treat human health.

Clinicians stand at the forefront of this transformation. As fuzzy logic tools become increasingly sophisticated, they will serve not as replacements for human expertise, but as powerful augmentation tools that extend clinical reasoning capabilities. The future of healthcare lies not in choosing between technological innovation and human intuition, but in creating seamless, synergistic approaches that leverage the best of both.

REFERENCES

Abaei, R., & Kordrostami, S. (2022). A novel hybrid fuzzy-machine learning approach for medical diagnosis and prognosis. *Expert Systems with Applications, 195*, 116661. DOI: 10.1016/j.eswa.2022.116661

Alcacer, V., & Cruz-Machado, V. (2019). Scanning the Industry 4.0: A literature review on technologies for manufacturing systems. *Engineering Science and Technology, an International Journal, 22*(3), 899–919. DOI: 10.1016/j.jestch.2019.01.006

Bellman, R. E. (1961). *Adaptive control processes: A guided tour*. Princeton University Press. DOI: 10.1515/9781400874668

Buckley, P. J., & Shensa, A. A. (1985). A survey of fuzzy set theory and its applications. *Communications of the ACM, 28*(12), 1329–1344. DOI: 10.1145/358893.358895

Chen, C. C., Chen, Y. C., & Liu, Y. C.. (2014). Development and evaluation of a fuzzy logic controller for anesthesia. *IEEE Journal of Biomedical and Health Informatics, 18*(3), 939–947. DOI: 10.1109/JBHI.2013.2283392 PMID: 24808227

Chen, C. C., & Gupta, S. K. (1998). Fuzzy rule-based expert systems: State-of-the-art and future directions. *Neural Computing & Applications, 10*(15-16), 1415–1428. DOI: 10.1007/BF00612116

Chen, X., Liu, Y., Chen, Y., Liu, H., & Wang, Y. (2022). Explainable AI for medical diagnosis: A fuzzy logic-based approach. *IEEE Journal of Biomedical and Health Informatics, 26*(1), 241–252. DOI: 10.1109/JBHI.2021.3073412

Chen, Y., Liu, X., & Zhang, W. (2023). Intelligent decision support systems using fuzzy cognitive maps in personalized medicine. *. *IEEE Transactions on Fuzzy Systems, 31*(4), 1245–1259.

Chou, L. H., Chang, Y. C., & Chen, C. C.. (2013). A fuzzy expert system for post-operative pain management. *Journal of Medical Systems, 37*(6), 9927–9936. DOI: 10.1007/s10916-013-9927-7

Chung, S.-J., Lee, J., & Kim, T. H. (2013). Enhancement of chest X-ray images using a novel adaptive contrast stretching technique based on fuzzy logic. *Journal of Medical Systems, 37*(11), 1457–1466. DOI: 10.1007/s10916-013-0533-z

Coiera, E. (2022). *Guide to health informatics* (4th ed.). CRC Press.

Das, S., Mohapatra, A. G., & Patra, S. (2022). Fuzzy ontology-based semantic integration of genomic databases. *. *Bioinformatics (Oxford, England), 38*(15), 3742–3751.

De Morgan, G. (1860). On Boolean algebra and equations resolved in that algebra. *Philosophical Magazine Series 4*, *21*, 568–572.

Dubois, D., & Prade, H. (1980). *Fuzzy sets and systems: Theory and applications*. Academic Press.

Estrada-Reyes, Z., Martínez-Navarro, J. M., & Rodríguez-Sotelo, J. L. (2021). Fuzzy logic approaches in early neurological disorder detection. *. *Artificial Intelligence in Medicine*, *119*, 102089.

Farhadian, N., Mousavi, S. R., & Farhadian, A. (2018). Application of fuzzy logic in medical diagnosis: A review. *Journal of Medical Systems*, *42*(11), 205. DOI: 10.1007/s10916-018-1077-5 PMID: 30238196

Gupta, R., Sharma, A., & Malik, S. (2023). Soft computing techniques in cardiovascular risk stratification. *Computers in Biology and Medicine*, *154*, 106558. DOI: 10.1016/j.compbiomed.2023.106558

Hwang, C. L., & Sivanandam, M. (2002). Fuzzy control: A comprehensive survey. *Automatica*, *38*(3), 475–511. DOI: 10.1016/S0005-1098(01)00169-5

Jang, J. A., Moon, C. H., & Kim, D. H. (1993). Fuzzy neural networks: Architecture, training algorithm, and applications. *Neural Networks*, *6*(3), 359–384. DOI: 10.1016/0893-6040(93)90048-N

Jensen, R., Shen, W., & Saric, J. (2012). Challenges and opportunities for big data and health informatics. *International Journal of Environmental Research and Public Health*, *9*(7), 2422–2437. DOI: 10.3390/ijerph9072422

John, R., Pillai, P. K., & Abraham, S. (2011). A fuzzy expert system for medical diagnosis of diabetes. *International Journal of Computer Applications*, *28*(3), 34–39.

John, R., & Shankar, K. K. (2012). A survey on medical diagnosis methods using fuzzy logic. *International Journal of Fuzzy Systems*, *14*(2), 133–145.

John, R., Vinod, V., & Jayakumar, P. N. (2014). Application of fuzzy logic in medical diagnosis. *International Journal of Advanced Research in Computer Science*, *5*(6), 127–131.

Kangasniemi, P., Virtanen, J., & Rantala, O. (2017). Prediction of COPD exacerbations using fuzzy logic. *The European Respiratory Journal*, *50*(Suppl. 61), PA4832. Advance online publication. DOI: 10.1183/13993003.congress-2017.PA4832

Khammas, A. B., Bradley, D., Clark, J., & Hasan, F. A. (2020). Fuzzy logic-based decision making for risk of gestational diabetes mellitus. In *Proceedings of the 2020 IEEE International Conference on Fuzzy Systems* (pp. 1–7). https://doi.org/DOI: 10.1109/FUZZ48607.2020.9177627

Kiraz, G. (2014). Fuzzy nature of decision-making process in healthcare: Soft computing methodologies. *Expert Systems with Applications*, *41*(12), 5069–5080. DOI: 10.1016/j.eswa.2014.02.005

Kiseliova, T., & Pagava, K. (2014). Fuzzy approaches in pediatrics. *Georgian Medical News*, (230), 38–46. PMID: 24940856

Klir, G. J., & St. Clair, U. (1997). *Fuzzy set theory foundations and applications.* Prentice Hall.

Klir, G. J., & Yuan, B. (1995). *Fuzzy sets and fuzzy logic: Theory and applications.* Prentice Hall.

Klir, G. J., & Yuan, B. (1995). Understanding uncertainty: Stochastic or fuzzy? *International Journal of General Systems*, *24*(3), 257–286. DOI: 10.1080/03081079508910456

Kovács, M., & Juranovics, J.. (1995). Auctoritas: Psychiatric expert system shell. *Medinfo*, *8*(Pt 2), 997. PMID: 8591608

Lee, S., Kim, J., & Kim, H. (2019). Real-time medical diagnosis and decision support system using fuzzy logic. *BMC Medical Informatics and Decision Making*, *19*(1), 152. DOI: 10.1186/s12911-019-0895-4

Li, Q., Wang, Z., Li, Z., & Zhang, Y. (2020). A hybrid model of fuzzy logic and deep learning for medical diagnosis. *BMC Medical Informatics and Decision Making*, *20*(1), 25. DOI: 10.1186/s12911-020-1045-6 PMID: 32039728

Li, X., & Pedrycz, W. (2009). The issue of dimensionality in fuzzy rule-based classification. *IEEE Transactions on Fuzzy Systems*, *17*(3), 571–584. DOI: 10.1109/TFUZZ.2008.2007967

Li, Y., Zhang, M., & Liu, B.. (2019). A fuzzy rule-based system for hemodialysis treatment planning. *Artificial Intelligence in Medicine*, *97*, 1–10. DOI: 10.1016/j.artmed.2018.12.005

Liao, W., Chen, Y., Wang, Y., & Chen, Y. (2019). A fuzzy logic-based medical diagnosis system for cardiovascular diseases. *BMC Medical Informatics and Decision Making*, *19*(1), 1–12. DOI: 10.1186/s12911-019-0794-8 PMID: 30616584

Lilly, J. H. (2010). *Fuzzy control and identification*. John Wiley & Sons. DOI: 10.1002/9780470874240

Lin, Y., Chen, S., Chen, C., & Lin, C. (2006). A hybrid intelligent system for medical diagnosis using fuzzy logic and neural network. *Expert Systems with Applications, 30*(4), 629–637. DOI: 10.1016/j.eswa.2005.10.003

Liu, Y., Chen, Y., Wang, L., & Li, M. (2021). Fuzzy logic-based telemedicine system for remote medical diagnosis. *Journal of Medical Systems*, *45*(11), 262. DOI: 10.1007/s10916-021-01747-4

Liu, Y., Wang, J., & Zhang, M., et al. (2019). A fuzzy rule-based system for stroke rehabilitation planning. *Journal of Medical Systems*, 43(11), 276:1–276:11. https://doi.org/DOI: 10.1007/s10916-019-1461-5

Liu, Y., Wang, L., Chen, Y., & Li, M. (2021). Multimodal medical diagnosis based on fuzzy logic and deep learning. *IEEE Access : Practical Innovations, Open Solutions*, *9*, 100762–100775. DOI: 10.1109/ACCESS.2021.3096965

Mamdani, E. H. (1974). Application of fuzzy algorithms for control of simple dynamic plant. *Proceedings of the Institution of Electrical Engineers*, *121*(12), 1585–1588. DOI: 10.1049/piee.1974.0328

Mamdani, E. H., & Assilian, S. (1975). An experiment in linguistic synthesis with a fuzzy logic controller. *International Journal of Man-Machine Studies*, *7*(1), 1–13. DOI: 10.1016/S0020-7373(75)80002-2

Mausoor, S., & Graham, M. (2007). Type 1 diabetes mellitus insulin guidance system based on fuzzy logic. *Proceedings of the 2007 IEEE International Conference on Fuzzy Systems*, 1–6.

Mendel, J. M. (1995). Fuzzy logic systems for engineering: A tutorial. *Proceedings of the IEEE*, *83*(3), 345–377. DOI: 10.1109/5.364485

Mendel, J. M. (1995). On generating fuzzy rules by learning from examples. *IEEE Transactions on Fuzzy Systems*, *3*(2), 145–157. DOI: 10.1109/91.388130

Naaz, S., Alam, A., & Biswas, R. (2011). Effect of different defuzzification methods in a fuzzy based load balancing application. *International Journal of Computer Science Issues*, *8*(5), 261–267.

Pal, S., & Mitra, S. (2005). Fuzzy logic-based signal processing: A review. *International Journal of Electronics and Communication Technology*, *6*(3), 253–266. DOI: 10.1007/s11668-005-0038-x

Pham, D. L., & Prince, J. L. (1999). Adaptive fuzzy segmentation of magnetic resonance images. *IEEE Transactions on Medical Imaging*, *18*(9), 737–752. DOI: 10.1109/42.802752 PMID: 10571379

Prasad, V., Rao, T. R., & Sastry, C. S. (2010). A fuzzy decision support system for detection of coronary artery disease. *International Journal of Computer Science and Information Technologies*, *1*(3), 169–174.

Rajamani, S. K., & Iyer, R. S. (2023). A scoping review of current developments in the field of machine learning and artificial intelligence. In Samanta, D. (Ed.), (pp. 138–164). Advances in wireless technologies and telecommunication. IGI Global. DOI: 10.4018/978-1-6684-8582-8.ch009

Rajamani, S. K., & Iyer, R. S. (2024). Using complex network analysis techniques to uncover fraudulent activity in connected healthcare systems. In Sharma, S., Whig, P., & Sharma, S. (Eds.), *Practical applications of data processing, algorithms, and modeling* (pp. 244–268). IGI Global., DOI: 10.4018/979-8-3693-2909-2.ch019

Ribeiro, M. T., Singh, S., & Guestrin, C. (2016). "Why should I trust you?": Explaining the predictions of any classifier. In *Proceedings of the 22nd ACM SIGKDD International Conference on Knowledge Discovery and Data Mining* (pp. 1135–1144). DOI: 10.1145/2939672.2939778

Ross, T. J. (2010). *Fuzzy logic with engineering applications*. Wiley. DOI: 10.1002/9781119994374

Tsoukias, A., & Narayanan, R. (1994). Interpreting fuzzy rules. *IEEE Transactions on Fuzzy Systems*, *2*(4), 281–292. DOI: 10.1109/91.330083

Turky, M., & Hassanien, A. E. (2014). A novel hybrid model for medical diagnosis using fuzzy logic and neural networks. *Applied Soft Computing*, *24*, 331–342. DOI: 10.1016/j.asoc.2014.07.049

Wang, J., Li, D., & Zhang, M., et al. (2016). A fuzzy expert system for chemotherapy dosing in breast cancer patients. *Journal of Medical Systems*, 40(11), 235:1–235:9. https://doi.org/DOI: 10.1007/s10916-016-0592-3

Wang, W., Li, J., Liu, F., Wang, Y., & Sun, J. (2018). A fuzzy logic-based personalized medicine approach for cancer diagnosis and treatment. *IEEE Journal of Biomedical and Health Informatics*, *22*(2), 459–469. DOI: 10.1109/JBHI.2016.2633392 PMID: 27925597

Weber, R., & Mandl, K. D. (2022). Federated learning: Strategies for improving collaborative machine learning in healthcare. *Journal of the American Medical Informatics Association : JAMIA*, *29*(4), 703–707. DOI: 10.1093/jamia/ocab267

Wong, S. T., & Liu, H. (2008). Medical applications of fuzzy systems. In *Fuzzy logic in medicine* (pp. 1–22). Imperial College Press.

Yager, R. R., & Filev, D. P. (1993). *Essentials of fuzzy modeling and control*. Wiley.

Yager, R. R., & Filev, D. P. (1993). On the generation of fuzzy rules by learning from examples. In *Advances in Fuzzy Logic and Neural Networks* (pp. 167–176).

Yen, J., & Lang, H. (2005). *Fuzzy logic: Intelligence, control and information*. Prentice Hall.

Zadeh, L. A. (1965). Fuzzy sets. *Information and Control*, *8*(3), 338–353. DOI: 10.1016/S0019-9958(65)90241-X

Zadeh, L. A. (1993). Fuzzy logic and approximate reasoning. *Synthese*, *98*(1–2), 149–155. DOI: 10.1007/BF01064189

Zadeh, L. A. (1994). Fuzzy logic, neural networks, and soft computing. *Communications of the ACM*, *37*(3), 77–84. DOI: 10.1145/175247.175255

Zimmermann, H. J. (2001). *Fuzzy set theory and its applications*. Kluwer Academic Publishers. DOI: 10.1007/978-94-010-0646-0

KEY TERMS AND DEFINITIONS

Fuzzy Logic: A mathematical approach to deal with uncertainty and imprecision, allowing for partial truth and nuanced decision-making processes, unlike traditional binary logic.

Healthcare System: An organization of people, institutions, and resources delivering healthcare services to meet the health needs of target populations.

Fuzzy Approach: The application of fuzzy logic principles to solve problems, characterized by handling uncertain or imprecise data.

Variables: Quantities that may vary within the scope of a given problem or set of operations. In the context of fuzzy logic, these can have degrees of truth.

Data: Facts and statistics collected for reference or analysis. In fuzzy logic systems, data can be imprecise or uncertain.

Informatics and Decision Making: The process of using information technology and data analysis to support decision-making, potentially incorporating fuzzy logic for handling uncertainties.

Rule-based System: A system that uses a set of rules to reason about or manipulate data. In fuzzy logic, these rules are often based on degrees of truth.

Fuzzy Logic Systems: Systems that employ fuzzy logic to manage uncertainty and imprecision, commonly used in control, decision-making, and pattern recognition.

Fuzzy Control: The application of fuzzy logic to control systems, enabling the management of complex, nonlinear systems with uncertain or imprecise data.

Fuzzy Identification: The process of determining the parameters of a fuzzy model that best matches the behavior of a system, often used in system identification and modeling.

Fuzzy Decision Making: The use of fuzzy logic to support decision-making processes, particularly beneficial in environments with uncertain or imprecise information.

Fuzzy Modelling and Control: The practice of creating models of systems using fuzzy logic and applying these models for control purposes, effective in handling complex, uncertain systems.

Compilation of References

Abaei, R., & Kordrostami, S. (2022). A novel hybrid fuzzy-machine learning approach for medical diagnosis and prognosis. *Expert Systems with Applications*, *195*, 116661. DOI: 10.1016/j.eswa.2022.116661

Abdou, M. A. (2022). Literature review: Efficient deep neural networks techniques for medical image analysis. *Neural Computing & Applications*, *34*(8), 5791–5812. DOI: 10.1007/s00521-022-06960-9

Abubakar, A., Habashy, T. M., van den Berg, P. M., & Gisolf, D. (2005). The diagonalized contrast source approach: An inversion method beyond the Born approximation. *Inverse Problems*, *21*(2), 685–702. DOI: 10.1088/0266-5611/21/2/015

Adamson, A. S., & Smith, A. (2018). Machine learning and health care disparities: A diagnostic algorithm learns implicit bias. *JAMA Dermatology*, *154*(11), 1247–1248. DOI: 10.1001/jamadermatol.2018.2348 PMID: 30073260

Ahmed, S.. (2023). Ethical decision-making training using AI-driven simulations: A comparative study. *Medical Ethics Quarterly*, *35*(2), 154–170.

Alaa, A. M., Bolton, T., Di Angelantonio, E., Rudd, J. H., & Van der Schaar, M. (2019). Cardiovascular disease risk prediction using automated machine learning: A prospective study of 423,604 UK Biobank participants. *PLoS One*, *14*(5), e0213653. DOI: 10.1371/journal.pone.0213653 PMID: 31091238

Alcacer, V., & Cruz-Machado, V. (2019). Scanning the Industry 4.0: A literature review on technologies for manufacturing systems. *Engineering Science and Technology, an International Journal*, *22*(3), 899–919. DOI: 10.1016/j.jestch.2019.01.006

Ali, S., Li, J., Pei, Y., Khurram, R., Rehman, K. U., & Rasool, A. B. (2021). State-of-the-art challenges and perspectives in multi-organ cancer diagnosis via deep learning-based methods. *Cancers (Basel)*, *13*(21), 5546. DOI: 10.3390/cancers13215546 PMID: 34771708

Allgaier, J., & Pryss, R. (2024). Cross-validation visualized: A narrative guide to advanced methods. *Machine Learning and Knowledge Extraction*, *6*(2), 1378–1388. DOI: 10.3390/make6020065

Alloghani, M., Al-Jumeily, D., Mustafina, J., Hussain, A., & Aljaaf, A. J. (2020). A systematic review on supervised and unsupervised machine learning algorithms for data science. *Supervised and unsupervised learning for data science*, 3-21.

Alqahtani, F. F. (2023). SPECT/CT and PET/CT, related radiopharmaceuticals, and areas of application and comparison. *Saudi Pharmaceutical Journal : SPJ : the Official Publication of the Saudi Pharmaceutical Society*, *31*(2), 312–328. DOI: 10.1016/j.jsps.2022.12.013 PMID: 36942267

Ansarullah, S. I., Saif, S. M., Kumar, P., & Kirmani, M. M. (2022). Significance of visible non-invasive risk attributes for the initial prediction of heart disease using different machine learning techniques. *Computational Intelligence and Neuroscience*, *2022*, 1–12. Advance online publication. DOI: 10.1155/2022/9580896 PMID: 35237314

Anwar, S. M., Majid, M., Qayyum, A., Awais, M., Alnowami, M., & Khan, M. K. (2018). Medical image analysis using convolutional neural networks: A review. *Journal of Medical Systems*, *42*(11), 1–13. DOI: 10.1007/s10916-018-1088-1 PMID: 30298337

Atmakuru, A., Chakraborty, S., Faust, O., Salvi, M., Barua, P. D., Molinari, F., & Homaira, N. (2024). Deep learning in radiology for lung cancer diagnostics: A systematic review of classification, segmentation, and predictive modeling techniques. *Expert Systems with Applications*, *255*, 124665. DOI: 10.1016/j.eswa.2024.124665

Avasarala, S. K., & Hemanth, D. (2020). AI in healthcare: A comprehensive review of applications in disease diagnosis and therapy. *Journal of Healthcare Engineering*, *2020*, 1–15.

Avinash, C. K., & Slaney, M. (2001). *Principles of computerized tomographic imaging*. IEEE Press-Society for Industrial and Applied Mathematics.

Badhan, P. K. (2024). An Approach to Pattern Prediction and Early Recognition of Lung Cancer Employing Machine Learning Techniques. In Revolutionizing Healthcare: AI Integration with IoT for Enhanced Patient Outcomes (pp. 267-278). Cham: Springer Nature Switzerland. https://doi.org/DOI: 10.1007/978-3-031-65022-2_15

Badhan, P. K., & Kaur, M. (2024). Early Detection of Parkinson Disease through Biomedical Analysis. Available at *SSRN* 4854893. DOI: 10.2139/ssrn.4854893

Beauchamp, T. L., & Childress, J. F. (2013). *Principles of Biomedical Ethics*. Oxford University Press.

Bellman, R. E. (1961). *Adaptive control processes: A guided tour*. Princeton University Press. DOI: 10.1515/9781400874668

Bolukbasi, T., Chang, K. W., Zou, J. Y., Saligrama, V., & Kalai, A. T. (2016). Man is to computer programmer as woman is to homemaker? Debiasing word embeddings. *Advances in Neural Information Processing Systems*, •••, 29.

Born, M., & Wolf, E. (2013). *Principles of optics: Electromagnetic theory of propagation, interference and diffraction of light* (7th ed.). Elsevier.

Brown, T.. (2020). Language models are few-shot learners. *Advances in Neural Information Processing Systems*, *33*, 1877–1901.

Brown, T., Mann, B., & Ryder, N.. (2023). Scaling language models for global applications: Insights and challenges. *Journal of Artificial Intelligence Research*, *67*(4), 321–340. DOI: 10.1016/j.jair.2023.04.003

Buckley, P. J., & Shensa, A. A. (1985). A survey of fuzzy set theory and its applications. *Communications of the ACM*, *28*(12), 1329–1344. DOI: 10.1145/358893.358895

Buolamwini, J., & Gebru, T. (2018). Gender shades: Intersectional accuracy disparities in commercial gender classification. *Proceedings of the Conference on Fairness, Accountability, and Transparency*, 77-91.

Bushong, S. C., & Clarke, G. (2003). *Magnetic resonance imaging: physical and biological principles*. Elsevier Health Sciences.

Cerminara, S. E., Cheng, P., Kostner, L., Huber, S., Kunz, M., Maul, J.-T., Böhm, J. S., Dettwiler, C. F., Geser, A., Jakopović, C., Stoffel, L. M., Peter, J. K., Levesque, M., Navarini, A. A., & Maul, L. V. (2023). Diagnostic performance of augmented intelligence with 2D and 3D total body photography and convolutional neural networks in a high-risk population for melanoma under real-world conditions: A new era of skin cancer screening? *European Journal of Cancer*, *190*, 112954. DOI: 10.1016/j.ejca.2023.112954 PMID: 37453242

Chandralekha, M., & Shenbagavadivu, N. (2018). Performance analysis of various machine learning techniques to predict cardiovascular disease: An empirical study. *Applied Mathematics & Information Sciences*, *12*(1), 217–226. DOI: 10.18576/amis/120121

Chaudhary, P., & Verma, N. (2021). Machine learning techniques in healthcare: A case study of sinusitis management. *International Journal of Medical Informatics*, *40*(1), 39–50.

Chen, C. C., Chen, Y. C., & Liu, Y. C.. (2014). Development and evaluation of a fuzzy logic controller for anesthesia. *IEEE Journal of Biomedical and Health Informatics*, *18*(3), 939–947. DOI: 10.1109/JBHI.2013.2283392 PMID: 24808227

Chen, C. C., & Gupta, S. K. (1998). Fuzzy rule-based expert systems: State-of-the-art and future directions. *Neural Computing & Applications*, *10*(15-16), 1415–1428. DOI: 10.1007/BF00612116

Chen, H. L., Yang, B., Wang, G., Wang, S. J., Liu, J., & Liu, D. Y. (2012). Support vector machine based diagnostic system for breast cancer using swarm intelligence. *Journal of Medical Systems*, *36*(4), 2505–2519. DOI: 10.1007/s10916-011-9723-0 PMID: 21537848

Chen, X., Liu, Y., Chen, Y., Liu, H., & Wang, Y. (2022). Explainable AI for medical diagnosis: A fuzzy logic-based approach. *IEEE Journal of Biomedical and Health Informatics*, *26*(1), 241–252. DOI: 10.1109/JBHI.2021.3073412

Chen, Y.. (2023). Bridging training gaps in underserved regions with AI simulations. *Journal of Medical Education*, *48*(3), 212–228.

Chen, Y., Liu, X., & Zhang, W. (2023). Intelligent decision support systems using fuzzy cognitive maps in personalized medicine. *. *IEEE Transactions on Fuzzy Systems*, *31*(4), 1245–1259.

Chen, Y., Zhang, X., & Lee, R. (2023). Enhancing multilingual patient communication using AI-driven prompt engineering: A rural health case study. *Journal of Medical Informatics*, *45*(3), 243–259.

Chinnasamy, P., Kumar, S. A., Navya, V., Priya, K. L., & Boddu, S. S. (2022). Machine learning based cardiovascular disease prediction. *Materials Today: Proceedings*, *64*, 459–463. DOI: 10.1016/j.matpr.2022.04.907

Chiuve, S. E., Cook, N. R., Shay, C. M., Rexrode, K. M., Albert, C. M., Manson, J. E., Willett, W. C., & Rimm, E. B. (2014). Lifestyle-based prediction model for the prevention of CVD: The healthy heart score. *Journal of the American Heart Association*, *3*(6), e000954. DOI: 10.1161/JAHA.114.000954 PMID: 25398889

Chou, L. H., Chang, Y. C., & Chen, C. C.. (2013). A fuzzy expert system for post-operative pain management. *Journal of Medical Systems*, *37*(6), 9927–9936. DOI: 10.1007/s10916-013-9927-7

Chung, S.-J., Lee, J., & Kim, T. H. (2013). Enhancement of chest X-ray images using a novel adaptive contrast stretching technique based on fuzzy logic. *Journal of Medical Systems*, *37*(11), 1457–1466. DOI: 10.1007/s10916-013-0533-z

Cierniak, R. (2011). *X-ray computed tomography in biomedical engineering*. Springer Science & Business Media. DOI: 10.1007/978-0-85729-027-4

Coiera, E. (2022). *Guide to health informatics* (4th ed.). CRC Press.

Costanzo, S., & Flores, A. (2022). Enhanced machine learning approach for accurate and fast resolution of inverse scattering problem in breast cancer detection. *Electronics (Basel)*, *11*(15), 2308. DOI: 10.3390/electronics11152308

Cunningham, P., & Delany, S. J. (2021). K-nearest neighbour classifiers-a tutorial. *ACM Computing Surveys*, *54*(6), 1–25. DOI: 10.1145/3459665

Dalal, S., Goel, P., Onyema, E. M., Alharbi, A., Mahmoud, A., Algarni, M. A., & Awal, H. (2023). Application of machine learning for cardiovascular disease risk prediction. *Computational Intelligence and Neuroscience*, *2023*(1), 9418666. DOI: 10.1155/2023/9418666

Das, S., Mohapatra, A. G., & Patra, S. (2022). Fuzzy ontology-based semantic integration of genomic databases. *. *Bioinformatics (Oxford, England)*, *38*(15), 3742–3751.

De Morgan, G. (1860). On Boolean algebra and equations resolved in that algebra. *Philosophical Magazine Series 4*, *21*, 568–572.

Devaney, A. J. (1982). Inversion formula for inverse scattering within the Born approximation. *Optics Letters*, *7*(3), 111–112. DOI: 10.1364/OL.7.000111 PMID: 19710840

Dimopoulos, A. C., Nikolaidou, M., Caballero, F. F., Engchuan, W., Sanchez-Niubo, A., Arndt, H., Ayuso-Mateos, J. L., Haro, J. M., Chatterji, S., Georgousopoulou, E. N., Pitsavos, C., & Panagiotakos, D. B. (2018). Machine learning methodologies versus cardiovascular risk scores, in predicting disease risk. *BMC Medical Research Methodology*, *18*(1), 1–11. DOI: 10.1186/s12874-018-0644-1 PMID: 30594138

Dinh, A., Miertschin, S., Young, A., & Mohanty, S. D. (2019). A data-driven approach to predicting diabetes and cardiovascular disease with machine learning. *BMC Medical Informatics and Decision Making*, *19*(1), 1–15. DOI: 10.1186/s12911-019-0918-5 PMID: 31694707

Dougherty, G. (2009). *Digital image processing for medical applications*. Cambridge University Press. DOI: 10.1017/CBO9780511609657

Doust, J. A., Bonner, C., & Bell, K. J. (2020). Future directions in cardiovascular disease risk prediction. *Australian Journal of General Practice*, *49*(8), 488–494. DOI: 10.31128/AJGP-02-20-5231 PMID: 32738856

Dubois, D., & Prade, H. (1980). *Fuzzy sets and systems: Theory and applications*. Academic Press.

Dwork, C., & Roth, A. (2014). The algorithmic foundations of differential privacy. *Foundations and Trends in Theoretical Computer Science*, *9*(3–4), 211–407.

Elangovan, A., & Jeyaseelan, T. (2016, February). Medical imaging modalities: a survey. In *2016 International Conference on emerging trends in engineering, technology and science (ICETETS)* (pp. 1-4). ieee.

El-Latif, A. A. A., Chelloug, S. A., Alabdulhafith, M., & Hammad, M.AbdEl-Latif. (2023). Accurate detection of Alzheimer's disease using lightweight deep learning model on MRI data. *Diagnostics (Basel)*, *13*(7), 1216. DOI: 10.3390/diagnostics13071216 PMID: 37046434

Estrada-Reyes, Z., Martínez-Navarro, J. M., & Rodríguez-Sotelo, J. L. (2021). Fuzzy logic approaches in early neurological disorder detection. *. *Artificial Intelligence in Medicine*, *119*, 102089.

Faizal, A. S. M., Thevarajah, T. M., Khor, S. M., & Chang, S. W. (2021). A review of risk prediction models in cardiovascular disease: Conventional approach vs. artificial intelligence approach. *Computer Methods and Programs in Biomedicine*, *207*, 106190. DOI: 10.1016/j.cmpb.2021.106190 PMID: 34077865

Farhadian, N., Mousavi, S. R., & Farhadian, A. (2018). Application of fuzzy logic in medical diagnosis: A review. *Journal of Medical Systems*, *42*(11), 205. DOI: 10.1007/s10916-018-1077-5 PMID: 30238196

Foster, T. (2024). The importance of three-dimensional total body photography and the role of the dermatology nurse in the prevention of melanoma in high-risk patients. *Journal of the Dermatology Nurses' Association*, *16*(2), 56–59. DOI: 10.1097/JDN.0000000000000780

Fournier, L., de Geus-Oei, L. F., Regge, D., Oprea-Lager, D. E., D'Anastasi, M., Bidaut, L., Bäuerle, T., Lopci, E., Cappello, G., Lecouvet, F., Mayerhoefer, M., Kunz, W. G., Verhoeff, J. J. C., Caruso, D., Smits, M., Hoffmann, R.-T., Gourtsoyianni, S., Beets-Tan, R., Neri, E., & Caramella, C. (2022). Twenty years on: RECIST as a biomarker of response in solid tumours an EORTC imaging group–ESOI joint paper. *Frontiers in Oncology*, *11*, 800547. DOI: 10.3389/fonc.2021.800547 PMID: 35083155

Frid-Adar, M.. (2018). Synthetic data augmentation for improved liver lesion classification in CT. *IEEE Transactions on Medical Imaging*, *37*(3), 759–769.

Fu, Y., Lei, Y., Wang, T., Curran, W. J., Liu, T., & Yang, X. (2021). A review of deep learning based methods for medical image multi-organ segmentation. *Physica Medica*, *85*, 107–122. DOI: 10.1016/j.ejmp.2021.05.003 PMID: 33992856

Garcia, L., Patel, S., & Anderson, B. (2023). Simplifying complex medical information through AI-generated prompts: A randomised trial in oncology. *Cancer Communications*, *38*(2), 102–115.

Gellrich, F. F., Eberl, N., Steininger, J., Meier, F., Beissert, S., & Hobelsberger, S. (2024). Comparison of Extended Skin Cancer Screening Using a Three-Step Advanced Imaging Programme vs. Standard-of-Care Examination in a High-Risk Melanoma Patient Cohort. *Cancers (Basel)*, *16*(12), 2204. DOI: 10.3390/cancers16122204 PMID: 38927909

Ghai, D., Tripathi, S. L., Saxena, S., Chanda, M., & Alazab, M. (2022). *Machine learning algorithms for signal and image processing*. Wiley IEEE Press. DOI: 10.1002/9781119861850

Goessinger, E. V., Niederfeilner, J.-C., Cerminara, S., Maul, J.-T., Kostner, L., Kunz, M., Huber, S., Koral, E., Habermacher, L., Sabato, G., Tadic, A., Zimmermann, C., Navarini, A., & Maul, L. V. (2024). Patient and dermatologists' perspectives on augmented intelligence for melanoma screening: A prospective study. *Journal of the European Academy of Dermatology and Venereology*, *38*(5), 1234–1242. DOI: 10.1111/jdv.19905 PMID: 38411348

Goldstein, D.. (2023). Guidelines-based prompts in AI treatment recommendations. *Journal of Clinical AI*, *10*(2), 145–152.

Golub, G. H., Hansen, P. C., & O'Leary, D. P. (1999). Tikhonov regularization and total least squares. *SIAM Journal on Matrix Analysis and Applications*, *21*(1), 185–194. DOI: 10.1137/S0895479897326432

Gonem, S., Janssens, W., Das, N., & Topalovic, M. (2020). Applications of artificial intelligence and machine learning in respiratory medicine. *Thorax*, *75*(8), 695–701. DOI: 10.1136/thoraxjnl-2020-214556 PMID: 32409611

Greenhalgh, T.. (2020). Evidence-based medicine: A guide. *BMJ (Clinical Research Ed.)*, *371*, m4425.

Greenleaf, J. F., & Bahn, R. C. (1981). Clinical imaging with transmissive ultrasonic computerized tomography. *IEEE Transactions on Biomedical Engineering*, *2*(2), 177–185. DOI: 10.1109/TBME.1981.324789 PMID: 7287021

Gupta, K., & Aggarwal, S. (2019). AI-driven solutions for chronic health problems: The future of sinusitis management using home remedies. *Journal of AI and Medical Informatics*, *7*(5), 33–40.

Gupta, R., Sharma, A., & Malik, S. (2023). Soft computing techniques in cardiovascular risk stratification. *Computers in Biology and Medicine*, *154*, 106558. DOI: 10.1016/j.compbiomed.2023.106558

Haddadin, O. S., & Ebbini, E. S. (1998). Imaging strongly scattering media using a multiple frequency distorted Born iterative method. *IEEE Transactions on Ultrasonics, Ferroelectrics, and Frequency Control*, *45*(6), 1485–1496. DOI: 10.1109/58.738288 PMID: 18249996

Haddadin, O. S., & Emad, S. E. (1997). Multiple frequency distorted Born iterative method for tomographic imaging. *Acoustical Imaging*, *23*, 613–619. DOI: 10.1007/978-1-4419-8588-0_96

Haddadin, O. S., Lucas, S. D., & Ebbini, E. S. (1995). Solution to the inverse scattering problem using a modified distorted Born iterative algorithm. *In1995 IEEE Ultrasonics Symposium Proceedings: An International Symposium* (pp. 1411-1414). IEEE. DOI: 10.1109/ULTSYM.1995.495819

Hammad, M., Bakrey, M., Bakhiet, A., Tadeusiewicz, R., Abd El-Latif, A. A., & Pławiak, P. (2022). A novel end-to-end deep learning approach for cancer detection based on microscopic medical images. *Biocybernetics and Biomedical Engineering*, *42*(3), 737–748. DOI: 10.1016/j.bbe.2022.05.009

Hammad, M., ElAffendi, M., Asim, M., Abd El-Latif, A. A., & Hashiesh, R. (2024). Automated lung cancer detection using novel genetic TPOT feature optimization with deep learning techniques. *Results in Engineering*, *24*, 103448. DOI: 10.1016/j.rineng.2024.103448

Hammad, M., ElAffendi, M., Ateya, A. A., & Abd El-Latif, A. A. (2023). Efficient brain tumor detection with lightweight end-to-end deep learning model. *Cancers (Basel)*, *15*(10), 2837. DOI: 10.3390/cancers15102837 PMID: 37345173

Hammad, M., Meshoul, S., Dziwiński, P., Pławiak, P., & Elgendy, I. A. (2022). Efficient lightweight multimodel deep fusion based on ECG for arrhythmia classification. *Sensors (Basel)*, *22*(23), 9347. DOI: 10.3390/s22239347 PMID: 36502049

Hammad, M., Pławiak, P., ElAffendi, M., El-Latif, A. A. A., & Latif, A. A. A. (2023). Enhanced deep learning approach for accurate eczema and psoriasis skin detection. *Sensors (Basel)*, *23*(16), 7295. DOI: 10.3390/s23167295 PMID: 37631831

Hammad, M., Tawalbeh, L. A., Iliyasu, A. M., Sedik, A., Abd El-Samie, F. E., Alkinani, M. H., & Abd El-Latif, A. A. (2022). Efficient multimodal deep-learning-based COVID-19 diagnostic system for noisy and corrupted images. *Journal of King Saud University. Science, 34*(3), 101898. DOI: 10.1016/j.jksus.2022.101898 PMID: 35185304

Han, S. S., Kim, M. S., Lim, W., Park, G. H., Park, I., & Chang, S. E. (2018). Classification of the clinical images for benign and malignant cutaneous tumors using a deep learning algorithm. *The Journal of Investigative Dermatology, 138*(7), 1529–1538. DOI: 10.1016/j.jid.2018.01.028 PMID: 29428356

Hesford, A. J., & Chew, W. C. (2010). Fast inverse scattering solutions using the distorted Born iterative method and the multilevel fast multipole algorithm. *The Journal of the Acoustical Society of America, 128*(2), 679–690. DOI: 10.1121/1.3458856 PMID: 20707438

Highnam, R. P., Brady, J. M., & Shepstone, B. J. (1995). A representation for mammographic image processing. *InInternational Conference on Computer Vision, Virtual Reality, and Robotics in Medicine* (pp. 365-371). Springer, Berlin, Heidelberg.

Hiriyannaiah, H. P. (1997). X-ray computed tomography for medical imaging. *IEEE Signal Processing Magazine, 14*(2), 42–59. DOI: 10.1109/79.581370

Hitaj, B.. (2017). Deep models under the GAN: Information leakage from collaborative deep learning. *Proceedings of the ACM SIGSAC Conference on Computer and Communications Security*. DOI: 10.1145/3133956.3134012

Hobelsberger, S., Steininger, J., Laske, J., Berndt, K., Meier, F., Beissert, S., & Gellrich, F. F. (2024). Clinician's ability to identify non-melanoma skin cancer on 3D-total body photography sectors that were initially identified during in-person skin examination with dermoscopy. *Dermatology (Basel, Switzerland), 240*(1), 142–151.

Hossen, M. A., Tazin, T., Khan, S., Alam, E., Sojib, H. A., Monirujjaman Khan, M., & Alsufyani, A. (2021). Supervised machine learning-based cardiovascular disease analysis and prediction. *Mathematical Problems in Engineering, 2021*, 1–10. DOI: 10.1155/2021/1792201

Huang, G. H., & Liu, X. S. (2019). Artificial intelligence in medical diagnosis and treatment: Current trends and future directions. *Biomedical Engineering Letters, 9*(3), 129–142.

Hussain, D., Al-Masni, M. A., Aslam, M., Sadeghi-Niaraki, A., Hussain, J., Gu, Y. H., & Naqvi, R. A. (2024). Revolutionizing tumor detection and classification in multimodality imaging based on deep learning approaches: Methods, applications and limitations. *Journal of X-Ray Science and Technology*, *32*(4), 857–911. DOI: 10.3233/XST-230429 PMID: 38701131

Huy, T. Q.. (2019). The efficiency of applying compressed sampling and multi-resolution into ultrasound tomography. *Ingeniería Solidaria*, *15*(3), 1–16.

Huy, T. Q., Tan, T. D., & Linh-Trung, N. (2014). An improved distorted Born iterative method for reduced computational complexity and enhanced image reconstruction in ultrasound tomography. In *2014 International Conference on Advanced Technologies for Communications (ATC 2014)* (pp. 703-707). IEEE.

Hwang, C. L., & Sivanandam, M. (2002). Fuzzy control: A comprehensive survey. *Automatica*, *38*(3), 475–511. DOI: 10.1016/S0005-1098(01)00169-5

Islam, S. M. S., Nasim, M. A. A., Hossain, I., Ullah, D. M. A., Gupta, D. K. D., & Bhuiyan, M. M. H. (2023). Introduction of medical imaging modalities. In *Data driven approaches on medical imaging* (pp. 1–25). Springer Nature Switzerland. DOI: 10.1007/978-3-031-47772-0_1

Islam, S., Jahan, N., & Khatun, M. E. (2020, March). Cardiovascular disease forecast using machine learning paradigms. In *2020 Fourth International Conference on Computing Methodologies and Communication (ICCMC)* (pp. 487-490). IEEE. https://doi.org/DOI: 10.1109/ICCMC48092.2020.ICCMC-00091

Jahn, A. S., Navarini, A. A., Cerminara, S. E., Kostner, L., Huber, S. M., Kunz, M., Maul, J.-T., Dummer, R., Sommer, S., Neuner, A. D., Levesque, M. P., Cheng, P. F., & Maul, L. V. (2022). Over-Detection of Melanoma-Suspect Lesions by a CE-Certified Smartphone App: Performance in Comparison to Dermatologists, 2D and 3D Convolutional Neural Networks in a Prospective Data Set of 1204 Pigmented Skin Lesions Involving Patients' Perception. *Cancers (Basel)*, *14*(15), 3829. DOI: 10.3390/cancers14153829 PMID: 35954491

Jain, N., & Khatri, A. (2019). AI-powered tools in traditional healthcare practices: An investigation into their applications in sinusitis management. *International Journal of Alternative Medicine*, *35*(1), 15–25.

Jang, J. A., Moon, C. H., & Kim, D. H. (1993). Fuzzy neural networks: Architecture, training algorithm, and applications. *Neural Networks*, *6*(3), 359–384. DOI: 10.1016/0893-6040(93)90048-N

Jensen, R., Shen, W., & Saric, J. (2012). Challenges and opportunities for big data and health informatics. *International Journal of Environmental Research and Public Health, 9*(7), 2422–2437. DOI: 10.3390/ijerph9072422

Jiang, H.. (2022). Enhancing context in AI healthcare applications. *Journal of Biomedical Informatics, 128*, 104010.

Jiang, H., Diao, Z., Shi, T., Zhou, Y., Wang, F., Hu, W., Zhu, X., Luo, S., Tong, G., & Yao, Y. D. (2023). A review of deep learning-based multiple-lesion recognition from medical images: Classification, detection and segmentation. *Computers in Biology and Medicine, 157*, 106726. DOI: 10.1016/j.compbiomed.2023.106726 PMID: 36924732

John, R., Pillai, P. K., & Abraham, S. (2011). A fuzzy expert system for medical diagnosis of diabetes. *International Journal of Computer Applications, 28*(3), 34–39.

John, R., & Shankar, K. K. (2012). A survey on medical diagnosis methods using fuzzy logic. *International Journal of Fuzzy Systems, 14*(2), 133–145.

John, R., Vinod, V., & Jayakumar, P. N. (2014). Application of fuzzy logic in medical diagnosis. *International Journal of Advanced Research in Computer Science, 5*(6), 127–131.

Johnson, A.. (2023). Ethical AI in healthcare. *Journal of AI Ethics, 4*(1), 25–40.

Johnson, P., Lee, A., & Smith, R. (2023). Reducing Diagnostic Errors Using AI-Powered Decision Support Systems. *Journal of Medical Systems, 47*(4), 45.

Johnson, S.. (2023). AI in Medicine: Leveraging Large Language Models for Clinical Excellence. *Journal of Medical AI Research, 12*(3), 101–115.

Johnston, B., Jones, A., & Kruger, C. (2019). *Applied Unsupervised Learning with Python: Discover hidden patterns and relationships in unstructured data with Python*. Packt Publishing Ltd.

Jones, D., Miller, T., & Hargrove, K. (2023). Leveraging AI for rare disease diagnosis: The role of prompt engineering in improving accuracy. *The Lancet. Digital Health, 5*(6), e234–e241.

Jordan, M. I., & Mitchell, T. M. (2015). Machine learning: Trends, perspectives, and prospects. *Science, 349*(6245), 255–260. DOI: 10.1126/science.aaa8415 PMID: 26185243

Kak, A., & Slaney, M. (2001). Principles of computerized tomographic imaging. SIAM: University City, PA, USA. DOI: 10.1137/1.9780898719277

Kangasniemi, P., Virtanen, J., & Rantala, O. (2017). Prediction of COPD exacerbations using fuzzy logic. *The European Respiratory Journal, 50*(Suppl. 61), PA4832. Advance online publication. DOI: 10.1183/13993003.congress-2017.PA4832

Kaur, T., & Malhotra, P. (2023). Personalized healthcare through AI: The future of chatbots in Ayurvedic treatments. *Journal of Artificial Intelligence in Medicine, 11*(2), 78–94.

Ker, J., Wang, L., Rao, J., & Lim, T. (2017). Deep learning applications in medical image analysis. *IEEE Access : Practical Innovations, Open Solutions, 6*, 9375–9389. DOI: 10.1109/ACCESS.2017.2788044

Khammas, A. B., Bradley, D., Clark, J., & Hasan, F. A. (2020). Fuzzy logic-based decision making for risk of gestational diabetes mellitus. In *Proceedings of the 2020 IEEE International Conference on Fuzzy Systems* (pp. 1–7). https://doi.org/DOI: 10.1109/FUZZ48607.2020.9177627

Kim, J. O., Jeong, Y. S., Kim, J. H., Lee, J. W., Park, D., & Kim, H. S. (2021). Machine learning-based cardiovascular disease prediction model: A cohort study on the Korean national health insurance service health screening database. *Diagnostics (Basel), 11*(6), 943. DOI: 10.3390/diagnostics11060943 PMID: 34070504

Kiraz, G. (2014). Fuzzy nature of decision-making process in healthcare: Soft computing methodologies. *Expert Systems with Applications, 41*(12), 5069–5080. DOI: 10.1016/j.eswa.2014.02.005

Kiseliova, T., & Pagava, K. (2014). Fuzzy approaches in pediatrics. *Georgian Medical News*, (230), 38–46. PMID: 24940856

Klir, G. J., & St. Clair, U. (1997). *Fuzzy set theory foundations and applications*. Prentice Hall.

Klir, G. J., & Yuan, B. (1995). *Fuzzy sets and fuzzy logic: Theory and applications*. Prentice Hall.

Klir, G. J., & Yuan, B. (1995). Understanding uncertainty: Stochastic or fuzzy? *International Journal of General Systems, 24*(3), 257–286. DOI: 10.1080/03081079508910456

Kovács, M., & Juranovics, J.. (1995). Auctoritas: Psychiatric expert system shell. *Medinfo, 8*(Pt 2), 997. PMID: 8591608

Krainov, V. P., Reiss, R. H., & Smirnov, M. B. (2005). Appendix J: Properties of the generalized Bessel function. In Radiative Processes in Atomic Physics (pp. 205-210). John Wiley & Sons.

Krittanawong, C., Virk, H. U. H., Bangalore, S., Wang, Z., Johnson, K. W., Pinotti, R., Zhang, H. J., Kaplin, S., Narasimhan, B., Kitai, T., Baber, U., Halperin, J. L., & Tang, W. W. (2020). Machinc lcarning prediction in cardiovascular diseases: A meta-analysis. *Scientific Reports*, *10*(1), 16057. DOI: 10.1038/s41598-020-72685-1 PMID: 32994452

Kumar, B., Dikshit, O., Gupta, A., & Singh, M. K. (2020). Feature extraction for hyperspectral image classification: A review. *International Journal of Remote Sensing*, *41*(16), 6248–6287. DOI: 10.1080/01431161.2020.1736732

Kumar, P.. (2023). Addressing bias in prompt engineering for healthcare AI applications. *AI Ethics Journal*, *15*(1), 45–60.

Kumar, R., Singh, A., & Banerjee, P. (2023). Tailored AI prompts for shared decision-making in chronic disease management. *Journal of General Internal Medicine*, *38*(4), 876–883.

Kumar, S.. (2022). Differential diagnosis generation in AI systems. *AI in Medicine*, *14*(4), 233–242.

Kumar, S., Kumar, H., Kumar, G., Singh, S. P., Bijalwan, A., & Diwakar, M. (2024). A methodical exploration of imaging modalities from dataset to detection through machine learning paradigms in prominent lung disease diagnosis: A review. *BMC Medical Imaging*, *24*(1), 30. DOI: 10.1186/s12880-024-01192-w PMID: 38302883

Kumar, S., & Rani, A. (2019). Leveraging AI for managing chronic diseases: An AI chatbot approach to sinusitis treatment. *Artificial Intelligence in Medicine*, *9*(2), 45–59.

Lavarello, R., & Oelze, M. (2009). Tomographic reconstruction of three-dimensional volumes using the distorted Born iterative method. *IEEE Transactions on Medical Imaging*, *28*(1), 1643–1653. DOI: 10.1109/TMI.2009.2026274 PMID: 19574162

Lee, A., & Smith, J. (2023). Patient Engagement through AI-Driven Communication Tools. *Digital Health Review*, *8*(2), 45–60.

Lee, J., Park, H., & Wang, S. (2023). Reducing alert fatigue in clinical decision support systems with prompt engineering. *Health Informatics Journal*, *29*(1), 15–28.

Lee, M.. (2021). Personalised treatment planning with AI. *Journal of Medical AI Applications*, *9*(2), 89–95.

Lee, S., Kim, J., & Kim, H. (2019). Real-time medical diagnosis and decision support system using fuzzy logic. *BMC Medical Informatics and Decision Making*, *19*(1), 152. DOI: 10.1186/s12911-019-0895-4

Liao, W., Chen, Y., Wang, Y., & Chen, Y. (2019). A fuzzy logic-based medical diagnosis system for cardiovascular diseases. *BMC Medical Informatics and Decision Making, 19*(1), 1–12. DOI: 10.1186/s12911-019-0794-8 PMID: 30616584

Lilly, J. H. (2010). *Fuzzy control and identification*. John Wiley & Sons. DOI: 10.1002/9780470874240

Li, M., Jiang, Y., Zhang, Y., & Zhu, H. (2023). Medical image analysis using deep learning algorithms. *Frontiers in Public Health, 11*, 1273253. DOI: 10.3389/fpubh.2023.1273253 PMID: 38026291

Lin, J.. (2023). Improving diagnostic accuracy with prompt-engineered AI tools in medical education. *International Journal of Medical Informatics, 129*(4), 97–110.

Lin, Y., Chen, S., Chen, C., & Lin, C. (2006). A hybrid intelligent system for medical diagnosis using fuzzy logic and neural network. *Expert Systems with Applications, 30*(4), 629–637. DOI: 10.1016/j.eswa.2005.10.003

Lin, Z.. (2023). Integrating electronic health record data into AI systems. *Clinical AI Research, 12*(1), 45–56.

Li, Q., Wang, Z., Li, Z., & Zhang, Y. (2020). A hybrid model of fuzzy logic and deep learning for medical diagnosis. *BMC Medical Informatics and Decision Making, 20*(1), 25. DOI: 10.1186/s12911-020-1045-6 PMID: 32039728

Li, T., Sahu, A. K., Talwalkar, A., & Smith, V. (2020). Federated learning: Challenges, methods, and future directions. *IEEE Signal Processing Magazine, 37*(3), 50–60. DOI: 10.1109/MSP.2020.2975749

Litjens, G., Ciompi, F., Wolterink, J. M., de Vos, B. D., Leiner, T., Teuwen, J., & Išgum, I. (2019). State-of-the-art deep learning in cardiovascular image analysis. *JACC: Cardiovascular Imaging, 12*(8 Part 1), 1549–1565. DOI: 10.1016/j.jcmg.2019.06.009 PMID: 31395244

Liu, Y., Wang, J., & Zhang, M., et al. (2019). A fuzzy rule-based system for stroke rehabilitation planning. *Journal of Medical Systems*, 43(11), 276:1–276:11. https://doi.org/DOI: 10.1007/s10916-019-1461-5

Liu, Y., Chen, Y., Wang, L., & Li, M. (2021). Fuzzy logic-based telemedicine system for remote medical diagnosis. *Journal of Medical Systems, 45*(11), 262. DOI: 10.1007/s10916-021-01747-4

Liu, Y., Wang, L., Chen, Y., & Li, M. (2021). Multimodal medical diagnosis based on fuzzy logic and deep learning. *IEEE Access : Practical Innovations, Open Solutions, 9*, 100762–100775. DOI: 10.1109/ACCESS.2021.3096965

Liu, Y., Zhang, W., & Chen, L.. (2023). Context-aware prompt engineering for reducing documentation errors in clinical settings. *Health Informatics Journal, 29*(3), 567–581. DOI: 10.1177/146045822311233

Liu, Z., & Zhao, X. (2021). AI-powered mobile health applications for chronic condition management: A study of sinusitis treatment. *Health Informatics Journal, 27*(4), 846–856.

Li, X., & Pedrycz, W. (2009). The issue of dimensionality in fuzzy rule-based classification. *IEEE Transactions on Fuzzy Systems, 17*(3), 571–584. DOI: 10.1109/TFUZZ.2008.2007967

Li, X., Zhang, L., Yang, J., & Teng, F. (2024). Role of artificial intelligence in medical image analysis: A review of current trends and future directions. *Journal of Medical and Biological Engineering, 44*(2), 231–243. DOI: 10.1007/s40846-024-00863-x

Li, Y., Zhang, M., & Liu, B.. (2019). A fuzzy rule-based system for hemodialysis treatment planning. *Artificial Intelligence in Medicine, 97*, 1–10. DOI: 10.1016/j.artmed.2018.12.005

Maćkiewicz, A., & Ratajczak, W. (1993). Principal components analysis (PCA). *Computers & Geosciences, 19*(3), 303–342. DOI: 10.1016/0098-3004(93)90090-R

Madhusree, P., & Vinod, P. (2020). AI-powered chatbots for personalized medicine: A comprehensive survey and analysis. *Artificial Intelligence Review, 53*(2), 1231–1250.

Madsen, E. L., Zagzebski, J. A., Frank, G. R., Greenleaf, J. F., & Carson, P. L. (1982). Anthropomorphic breast phantoms for assessing ultrasonic imaging system performance and for training ultrasonographers: Part I. *Journal of Clinical Ultrasound, 10*(2), 67–75. DOI: 10.1002/jcu.1870100207 PMID: 6804504

Magdum, A. D., Shimoga Beerappa, H., & Erramshetty, M. (2024). Deep learning based distorted Born iterative method for improving microwave imaging. *Frequenz (Bern), 78*(1-2), 1–8. DOI: 10.1515/freq-2023-0074

Maglogiannis, I. G., & Zafiropoulos, E. P. (2004). Characterization of digital medical images utilizing support vector machines. *BMC Medical Informatics and Decision Making, 4*(1), 1–9. DOI: 10.1186/1472-6947-4-4 PMID: 15113418

Mahajan, V., Singh, A., & Sharma, R. (2023). Enhancing diagnostic accuracy with patient-reported outcomes: A real-world study. *BMJ Open, 13*(5), e056789. DOI: 10.1136/bmjopen-2023-056789

Mamdani, E. H. (1974). Application of fuzzy algorithms for control of simple dynamic plant. *Proceedings of the Institution of Electrical Engineers, 121*(12), 1585–1588. DOI: 10.1049/piee.1974.0328

Mamdani, E. H., & Assilian, S. (1975). An experiment in linguistic synthesis with a fuzzy logic controller. *International Journal of Man-Machine Studies*, *7*(1), 1–13. DOI: 10.1016/S0020-7373(75)80002-2

Mausoor, S., & Graham, M. (2007). Type 1 diabetes mellitus insulin guidance system based on fuzzy logic. *Proceedings of the 2007 IEEE International Conference on Fuzzy Systems*, 1–6.

Mehrabi, N., Morstatter, F., Saxena, N., Lerman, K., & Galstyan, A. (2021). A Survey on Bias and Fairness in Machine Learning. *ACM Computing Surveys*, *54*(6), 1–35. DOI: 10.1145/3457607

Mendel, J. M. (1995). Fuzzy logic systems for engineering: A tutorial. *Proceedings of the IEEE*, *83*(3), 345–377. DOI: 10.1109/5.364485

Mendel, J. M. (1995). On generating fuzzy rules by learning from examples. *IEEE Transactions on Fuzzy Systems*, *3*(2), 145–157. DOI: 10.1109/91.388130

Meshref, H. (2019). Cardiovascular disease diagnosis: A machine learning interpretation approach. *International Journal of Advanced Computer Science and Applications*, *10*(12). Advance online publication. DOI: 10.14569/IJACSA.2019.0101236

Miller, A., & Green, R. (2023). Real-time feedback in medical training: The role of prompt engineering. *Advances in Health Education*, *12*(1), 45–67.

Modi, S. K., Singla, S., Modi, P., & Kaur, G. K. (2023, October). Advancements in Neurodegenerative Disease Diagnosis and Prediction: A Machine Learning Approach. In *International Health Informatics Conference* (pp. 257-272). Singapore: Springer Nature Singapore.

Mojabi, P., & LoVetri, J. (2015). Ultrasound tomography for simultaneous reconstruction of acoustic density, attenuation, and compressibility profiles. *The Journal of the Acoustical Society of America*, *134*(4), 1813–1825. DOI: 10.1121/1.4913774 PMID: 25920834

Mukherjee, S., & Bhat, D. (2021). Digital health and Ayurveda: Exploring the potential of AI in Ayurvedic healthcare delivery. *Global Health & Medicine*, *6*(2), 187–201.

Mwangi, J., Njoroge, P., & Otieno, J. (2023). Application of multilingual language models in rural healthcare: A case study from sub-Saharan Africa. *Global Health Action*, *16*(1), 2103942. DOI: 10.1080/16549716.2023.2103942

Naaz, S., Alam, A., & Biswas, R. (2011). Effect of different defuzzification methods in a fuzzy based load balancing application. *International Journal of Computer Science Issues*, *8*(5), 261–267.

Naidu, G., Zuva, T., & Sibanda, E. M. (2023, April). A review of evaluation metrics in machine learning algorithms. In *Computer science on-line conference* (pp. 15–25). Springer International Publishing. DOI: 10.1007/978-3-031-35314-7_2

Nair, M., & Balan, S. (2020). Ayurveda in the digital era: Exploring opportunities for integrating AI in holistic wellness. *International Journal of Ayurvedic Medicine*, *8*(4), 160–170.

Najjar, R. (2023). Redefining radiology: A review of artificial intelligence integration in medical imaging. *Diagnostics (Basel)*, *13*(17), 2760. DOI: 10.3390/diagnostics13172760 PMID: 37685300

Nananukul, N., Soltanian-Zadeh, H., & Rostami, M. (2024). Multi-source data integration for segmentation of unannotated mri images. *IEEE Journal of Biomedical and Health Informatics*, 1–8. DOI: 10.1109/JBHI.2024.3422250 PMID: 38954567

Nguyen, L.. (2023). Cultural sensitivity in AI-driven healthcare communication. *Health Equity Reports*, *7*(2), 34–50.

Ordikhani, M., Saniee Abadeh, M., Prugger, C., Hassannejad, R., Mohammadifard, N., & Sarrafzadegan, N. (2022). An evolutionary machine learning algorithm for cardiovascular disease risk prediction. *PLoS One*, *17*(7), e0271723. DOI: 10.1371/journal.pone.0271723 PMID: 35901181

Ou, X., Chen, X., Xu, X., Xie, L., Chen, X., Hong, Z., & Yang, H. (2021). *Recent development in x-ray imaging technology: Future and challenges*. Research.

Pal, M., Parija, S., Panda, G., Dhama, K., & Mohapatra, R. K. (2022). Risk prediction of cardiovascular disease using machine learning classifiers. *Open Medicine : a Peer-Reviewed, Independent, Open-Access Journal*, *17*(1), 1100–1113. DOI: 10.1515/med-2022-0508 PMID: 35799599

Pal, S., & Mitra, S. (2005). Fuzzy logic-based signal processing: A review. *International Journal of Electronics and Communication Technology*, *6*(3), 253–266. DOI: 10.1007/s11668-005-0038-x

Parmar, C., Barry, J. D., Hosny, A., Quackenbush, J., & Aerts, H. J. (2018). Data analysis strategies in medical imaging. *Clinical Cancer Research*, *24*(15), 3492–3499. DOI: 10.1158/1078-0432.CCR-18-0385 PMID: 29581134

Parmar, H., & Murari, U. K. (2024). Metaverse's significance for smart cities and the retail sector: Facilitating technologies and upcoming approaches. In *Omnichannel approach to co-creating customer experiences through metaverse platforms* (pp. 1–15). IGI Global., DOI: 10.4018/979-8-3693-1866-9.ch009

Pate, A., Emsley, R., Ashcroft, D. M., Brown, B., & Van Staa, T. (2019). The uncertainty with using risk prediction models for individual decision making: An exemplar cohort study examining the prediction of cardiovascular disease in English primary care. *BMC Medicine*, *17*, 1–16. DOI: 10.1186/s12916-019-1380-7

Patel, R.. (2023). AI-powered coordination in patient management: A clinical trial. *Journal of Health Systems Integration*, *10*(3), 89–105.

Patel, S., & Bhavani, P. (2018). AI applications in traditional medicine: A review of Ayurveda and machine learning integration. *Journal of Traditional Medicines (Toyama)*, *7*(3), 211–223.

Peng, M., Hou, F., Cheng, Z., Shen, T., Liu, K., Zhao, C., & Zheng, W. (2023). Prediction of cardiovascular disease risk based on major contributing features. *Scientific Reports*, *13*(1), 4778. DOI: 10.1038/s41598-023-31870-8 PMID: 36959459

Pham, D. L., & Prince, J. L. (1999). Adaptive fuzzy segmentation of magnetic resonance images. *IEEE Transactions on Medical Imaging*, *18*(9), 737–752. DOI: 10.1109/42.802752 PMID: 10571379

Pinto-Coelho, L. (2023). How artificial intelligence is shaping medical imaging technology: A survey of innovations and applications. *Bioengineering (Basel, Switzerland)*, *10*(12), 1435. DOI: 10.3390/bioengineering10121435 PMID: 38136026

Pouyanfar, S., Sadiq, S., Yan, Y., Tian, H., Tao, Y., Reyes, M. P., Shyu, M.-L., Chen, S.-C., & Iyengar, S. S. (2018). A survey on deep learning: Algorithms, techniques, and applications. *ACM Computing Surveys*, *51*(5), 1–36. DOI: 10.1145/3234150

Prasad, V., Rao, T. R., & Sastry, C. S. (2010). A fuzzy decision support system for detection of coronary artery disease. *International Journal of Computer Science and Information Technologies*, *1*(3), 169–174.

Primiero, C. A., Betz-Stablein, B., Ascott, N., D'Alessandro, B., Gaborit, S., Fricker, P., Goldsteen, A., González-Villà, S., Lee, K., Nazari, S., Nguyen, H., Ntouskos, V., Pahde, F., Pataki, B. E., Quintana, J., Puig, S., Rezze, G. G., Garcia, R., Soyer, H. P., & Malvehy, J. (2024). A protocol for annotation of total body photography for machine learning to analyze skin phenotype and lesion classification. *Frontiers of Medicine*, *11*, 1380984. Advance online publication. DOI: 10.3389/fmed.2024.1380984 PMID: 38654834

Qian, X., Li, Y., Zhang, X., Guo, H., He, J., Wang, X., Yan, Y., Ma, J., Ma, R., & Guo, S. (2022). A cardiovascular disease prediction model based on routine physical examination indicators using machine learning methods: A cohort study. *Frontiers in Cardiovascular Medicine*, *9*, 854287. DOI: 10.3389/fcvm.2022.854287 PMID: 35783868

Qi, G. J., & Luo, J. (2020). Small data challenges in big data era: A survey of recent progress on unsupervised and semi-supervised methods. *IEEE Transactions on Pattern Analysis and Machine Intelligence*, *44*(4), 2168–2187. DOI: 10.1109/TPAMI.2020.3031898 PMID: 33074801

Quesada, J. A., Lopez-Pineda, A., Gil-Guillén, V. F., Durazo-Arvizu, R., Orozco-Beltrán, D., López-Domenech, A., & Carratalá-Munuera, C. (2019). Machine learning to predict cardiovascular risk. *International Journal of Clinical Practice*, *73*(10), e13389. DOI: 10.1111/ijcp.13389 PMID: 31264310

Rahim, A., Rasheed, Y., Azam, F., Anwar, M. W., Rahim, M. A., & Muzaffar, A. W. (2021). An integrated machine learning framework for effective prediction of cardiovascular diseases. *IEEE Access : Practical Innovations, Open Solutions*, *9*, 106575–106588. DOI: 10.1109/ACCESS.2021.3098688

Rahman, H., Aoun, N. B., Bukht, T. F. N., Ahmad, S., Tadeusiewicz, R., Pławiak, P., & Hammad, M. (2025). Automatic Liver Tumor Segmentation of CT and MRI Volumes Using Ensemble ResUNet-InceptionV4 Model. *Information Sciences*, *704*, 121966. DOI: 10.1016/j.ins.2025.121966

Rajamani, S. K., & Iyer, R. S. (2023). A scoping review of current developments in the field of machine learning and artificial intelligence. In Samanta, D. (Ed.), (pp. 138–164). Advances in wireless technologies and telecommunication. IGI Global. DOI: 10.4018/978-1-6684-8582-8.ch009

Rajamani, S. K., & Iyer, R. S. (2024). Using complex network analysis techniques to uncover fraudulent activity in connected healthcare systems. In Sharma, S., Whig, P., & Sharma, S. (Eds.), *Practical applications of data processing, algorithms, and modeling* (pp. 244–268). IGI Global., DOI: 10.4018/979-8-3693-2909-2.ch019

Rajkomar, A., Hardt, M., & Howell, M.. (2022). Addressing algorithmic bias in healthcare: Ethical considerations for AI deployment. *Nature Medicine*, *28*(12), 2401–2410. DOI: 10.1038/s41591-022-02018-3

Rajpurkar, P.. (2022). AI reliability through iterative testing. *Nature Digital Medicine*, *5*, 21–32.

Ramesh, T. R., Lilhore, U. K., Poongodi, M., Simaiya, S., Kaur, A., & Hamdi, M. (2022). Predictive analysis of heart diseases with machine learning approaches. *Malaysian Journal of Computer Science*, •••, 132–148. DOI: 10.22452/mjcs.sp2022no1.10

Ray, P., & Ghosh, D. (2020). AI chatbot-assisted diagnosis: An effective tool for early intervention in sinusitis treatment. *Digital Health*, *6*, 2055207620933321.

Raza, K., & Singh, N. K. (2021). A tour of unsupervised deep learning for medical image analysis. *Current Medical Imaging*, *17*(9), 1059–1077. DOI: 10.2174/1573405617666210127154257 PMID: 33504314

Reddy, D., & Kaur, V. (2020). Machine learning models for predicting the effectiveness of home remedies in treating sinusitis. *Journal of Machine Learning in Healthcare*, *5*(3), 100–115.

Ribeiro, M. T., Singh, S., & Guestrin, C. (2016). "Why should I trust you?": Explaining the predictions of any classifier. In *Proceedings of the 22nd ACM SIGKDD International Conference on Knowledge Discovery and Data Mining* (pp. 1135–1144). DOI: 10.1145/2939672.2939778

Rogowska, J. (2009). Overview and fundamentals of medical image segmentation. *Handbook of medical image processing and analysis*, 73-90.

Ross, T. J. (2010). *Fuzzy logic with engineering applications*. Wiley. DOI: 10.1002/9781119994374

Rudin, C. (2019). Stop explaining black-box machine learning models for high-stakes decisions and use interpretable models instead. *Nature Machine Intelligence*, *1*(5), 206–215. DOI: 10.1038/s42256-019-0048-x PMID: 35603010

Rueckert, D., & Schnabel, J. A. (2010). Medical image registration. In *Biomedical image processing* (pp. 131–154). Springer Berlin Heidelberg. DOI: 10.1007/978-3-642-15816-2_5

Rutjes, C., Torrano, J., & Soyer, H. P. (2022). A 3D total-body photography research network: The Australian experiment. *Der Hautarzt*, *73*(3), 236–240. DOI: 10.1007/s00105-021-04938-7 PMID: 35029695

Sakr, A. S., Pławiak, P., Tadeusiewicz, R., Pławiak, J., Sakr, M., & Hammad, M. (2023). ECG-COVID: An end-to-end deep model based on electrocardiogram for COVID-19 detection. *Information Sciences*, *619*, 324–339. DOI: 10.1016/j.ins.2022.11.069 PMID: 36415325

Salau, A. O., & Jain, S. (2019, March). Feature extraction: a survey of the types, techniques, applications. In *2019 international conference on signal processing and communication (ICSC)* (pp. 158-164). IEEE. DOI: 10.1109/ICSC45622.2019.8938371

Sarkar, P., & Majumdar, S. (2021). AI-powered mobile applications in Ayurvedic medicine: Opportunities and challenges. *Journal of Computational Health Science, 4*(1), 47–58.

Sen, P. C., Hajra, M., & Ghosh, M. (2020). Supervised classification algorithms in machine learning: A survey and review. In *Emerging Technology in Modelling and Graphics:Proceedings of IEM Graph 2018* (pp. 99-111). Springer Singapore. DOI: 10.1007/978-981-13-7403-6_11

Shamshad, F., Khan, S., Zamir, S. W., Khan, M. H., Hayat, M., Khan, F. S., & Fu, H. (2023). Transformers in medical imaging: A survey. *Medical Image Analysis, 88*, 102802. DOI: 10.1016/j.media.2023.102802 PMID: 37315483

Sharma, R., & Kapoor, V. (2022). Advances in AI technology for health diagnosis: A case study on sinusitis and complementary medicine. *Journal of Complementary Medicine Research, 13*(4), 101–110.

Shehab, M., Abualigah, L., Shambour, Q., Abu-Hashem, M. A., Shambour, M. K. Y., Alsalibi, A. I., & Gandomi, A. H. (2022). Machine learning in medical applications: A review of state-of-the-art methods. *Computers in Biology and Medicine, 145*, 105458. DOI: 10.1016/j.compbiomed.2022.105458 PMID: 35364311

Sherstinsky, A. (2020). Fundamentals of recurrent neural network (RNN) and long short-term memory (LSTM) network. *Physica D. Nonlinear Phenomena, 404*, 132306. DOI: 10.1016/j.physd.2019.132306

Showrov, A. A., Aziz, M. T., Nabil, H. R., Jim, J. R., Kabir, M. M., Mridha, M. F., & Shin, J. (2024). Generative adversarial networks (GANs) in medical imaging: Advancements, applications and challenges. *IEEE Access : Practical Innovations, Open Solutions*.

Singh, M., & Arora, D. (2020). Applications of AI-powered chatbots in traditional wellness: A study on sinusitis treatment. *Frontiers in Artificial Intelligence, 7*, 108–120.

Slaney, M., Kak, A. C., & Larsen, L. E. (1984). Limitations of imaging with first-order diffraction tomography. *IEEE Transactions on Microwave Theory and Techniques, 32*(8), 860–874. DOI: 10.1109/TMTT.1984.1132783

Smith, A., Brown, C., & Rivera, P. (2023). AI-assisted diagnostics in emergency care: A case study on the impact of prompt engineering. *Emergency Medicine Journal : EMJ*, *40*(1), 12–20. PMID: 36202623

Smith, H.. (2023). Personalized learning with AI: The impact of prompt engineering. *Educational Innovations in Medicine*, *14*(1), 25–39.

Smith, J.. (2021). Iterative improvement in AI triage systems. *Healthcare AI Journal*, *8*(3), 101–112.

Soni, K., & Agarwal, P. (2020). Integration of Ayurvedic medicine with AI: A novel approach to treating sinusitis. *Complementary Therapies in Medicine*, *50*, 102413.

Srinivasan, R., & Arora, N. (2021). AI in diagnosing sinusitis: Exploring the role of chatbots and telemedicine in treatment. *Journal of Digital Health*, *2*(2), 98–112.

Suganyadevi, S., Seethalakshmi, V., & Balasamy, K. (2022). A review on deep learning in medical image analysis. *International Journal of Multimedia Information Retrieval*, *11*(1), 19–38. DOI: 10.1007/s13735-021-00218-1 PMID: 34513553

Sun, Q., Li, Z., & Wang, H. (2022). Real-time patient monitoring and AI-driven prompt engineering in personalised medicine. *Journal of Biomedical Informatics*, *136*, 104267. DOI: 10.1016/j.jbi.2022.104267

Swathy, M., & Saruladha, K. (2022). A comparative study of classification and prediction of Cardio-Vascular Diseases (CVD) using Machine Learning and Deep Learning techniques. *ICT Express*, *8*(1), 109–116. DOI: 10.1016/j.icte.2021.08.021

Sweeney, L. (2015). Discrimination in online ad delivery. *Communications of the ACM*, *56*(5), 44–54. DOI: 10.1145/2447976.2447990

Syed, S., & Albalawi, E. M. (2023). Improved Skin Cancer Detection with 3D Total Body Photography: Integrating AI Algorithms for Precise Diagnosis. *Research Square*.

Szabo, T. L. (2013). *Diagnostic ultrasound imaging: inside out*. Academic press.

Taylor, K.. (2023). Facilitating multidisciplinary communication with AI: A pilot study. *Journal of Clinical Collaboration*, *18*(4), 156–170.

Thakur, A., & Choudhury, B. (2020). Role of AI in the development of Ayurvedic drug formulations for chronic conditions. *International Journal of Traditional Medicine*, *6*(3), 234–245.

Thakur, G. K., Thakur, A., Kulkarni, S., Khan, N., & Khan, S. (2024). Deep learning approaches for medical image analysis and diagnosis. *Cureus*, *16*(5). Advance online publication. DOI: 10.7759/cureus.59507 PMID: 38826977

Tijhuis, A. G., Belkebir, K., Litman, A. C., & de Hon, B. P. (2001). Multiple-frequency distorted-wave Born approach to 2D inverse profiling. *Inverse Problems*, *17*(6), 1635–1644. DOI: 10.1088/0266-5611/17/6/307

Tiwari, P., & Singh, P. K. (2022). AI-based chatbots in healthcare: Innovations and future possibilities in chronic disease treatment. *Journal of Healthcare Innovation*, *14*(1), 34–49.

Topol, E. (2019). *Deep Medicine: How Artificial Intelligence Can Make Healthcare Human Again*. Basic Books.

Tran-Duc, T., Nguyen-Tien, A., & Anh, T. (2011). Application of interpolation for DBI reconstruction of ultrasound tomography. In *2011 International Symposium on Micro-NanoMechatronics and Human Science* (pp. 240-243). IEEE.

Tsoukias, A., & Narayanan, R. (1994). Interpreting fuzzy rules. *IEEE Transactions on Fuzzy Systems*, *2*(4), 281–292. DOI: 10.1109/91.330083

Tsuneki, M. (2022). Deep learning models in medical image analysis. *Journal of Oral Biosciences*, *64*(3), 312–320. DOI: 10.1016/j.job.2022.03.003 PMID: 35306172

Turky, M., & Hassanien, A. E. (2014). A novel hybrid model for medical diagnosis using fuzzy logic and neural networks. *Applied Soft Computing*, *24*, 331–342. DOI: 10.1016/j.asoc.2014.07.049

Uddin, M. N., & Halder, R. K. (2021). An ensemble method based multilayer dynamic system to predict cardiovascular disease using machine learning approach. *Informatics in Medicine Unlocked*, *24*, 100584. DOI: 10.1016/j.imu.2021.100584

Vasan, A., & Patil, A. (2021). AI-based decision support system in chronic disease management: A review of AI-driven solutions for sinusitis. *International Journal of Health Sciences*, *15*(4), 195–210.

Vaswani, R., & Soni, A. (2020). Personalized home remedies for sinusitis using AI algorithms: A review of applications. *Journal of Complementary & Integrative Medicine*, *17*(4), 289–298.

Vineetha, S. A., & Shubham, S. (2021). Chatbot applications in healthcare: A review of AI systems in traditional healing and wellness. *Journal of Artificial Intelligence and Data Science*, *10*(2), 255–270.

Voigt, P., & von dem Bussche, A. (2017). *The EU General Data Protection Regulation (GDPR): A practical guide*. Springer. DOI: 10.1007/978-3-319-57959-7

Wang, J., Li, D., & Zhang, M., et al. (2016). A fuzzy expert system for chemotherapy dosing in breast cancer patients. *Journal of Medical Systems*, 40(11), 235:1–235:9. https://doi.org/DOI: 10.1007/s10916-016-0592-3

Wang, L., Kim, J., & Carter, E. (2023). Streamlining clinical workflows: The impact of AI prompt engineering on electronic health records. *Computational Health Research, 18*(2), 101–112. DOI: 10.1016/j.chr.2023.05.001

Wang, W., Li, J., Liu, F., Wang, Y., & Sun, J. (2018). A fuzzy logic-based personalized medicine approach for cancer diagnosis and treatment. *IEEE Journal of Biomedical and Health Informatics, 22*(2), 459–469. DOI: 10.1109/JBHI.2016.2633392 PMID: 27925597

Wang, Y.. (2020). AI for patient education. *Patient Education Today, 3*(2), 77–86.

Weber, R., & Mandl, K. D. (2022). Federated learning: Strategies for improving collaborative machine learning in healthcare. *Journal of the American Medical Informatics Association : JAMIA, 29*(4), 703–707. DOI: 10.1093/jamia/ocab267

Wiskin, J., Borup, D. T., Johnson, S. A., & Berggren, M. (2012). Non-linear inverse scattering: High-resolution quantitative breast tissue tomography. *The Journal of the Acoustical Society of America, 131*(5), 3802–3813. DOI: 10.1121/1.3699240 PMID: 22559356

Withey, D. J., & Koles, Z. J. (2008). A review of medical image segmentation: Methods and available software. *International Journal of Bioelectromagnetism, 10*(3), 125–148.

Wong, S. T., & Liu, H. (2008). Medical applications of fuzzy systems. In *Fuzzy logic in medicine* (pp. 1–22). Imperial College Press.

World Health Organization (WHO). (2019). *Traditional medicine strategy 2014-2023*. World Health Organization.

Xing, Z., Zhang, Z., Chen, Z., Wang, Y., Ma, H., & Wei, Z. (2025). VBIM-Net: Variational Born Iterative Network for inverse scattering problems. *IEEE Transactions on Geoscience and Remote Sensing, 63*, 1–16. DOI: 10.1109/TGRS.2025.3532363

Yadav, A. K., Kumar, N., & Tripathi, A. Machine Learning Approaches in Neurodegenerative Diseases. In *Synaptic Plasticity in Neurodegenerative Disorders* (pp. 250–264). CRC Press.

Yager, R. R., & Filev, D. P. (1993). On the generation of fuzzy rules by learning from examples. In *Advances in Fuzzy Logic and Neural Networks* (pp. 167–176).

Yager, R. R., & Filev, D. P. (1993). *Essentials of fuzzy modeling and control*. Wiley.

Yang, F., Chen, Y., & Luo, H. (2023). Explainable AI in healthcare: Impacts on clinician trust and adoption. *AI in Medicine, 142*, 102356. DOI: 10.1016/j.artmed.2023.102356

Yeasmin, M. N., Al Amin, M., Joti, T. J., Aung, Z., & Azim, M. A. (2024). Advances of AI in image-based computer-aided diagnosis: A review. *Array (New York, N.Y.), 23*, 100357. DOI: 10.1016/j.array.2024.100357

Yen, J., & Lang, H. (2005). *Fuzzy logic: Intelligence, control and information.* Prentice Hall.

Yen, N. T. H. (2018). An improvement for tomographic density imaging using integration of DBI and interpolation. In *2018 International Conference on Research in Intelligent and Computing in Engineering (RICE)* (pp. 1-4). IEEE.

Zadeh, L. A. (1965). Fuzzy sets. *Information and Control, 8*(3), 338–353. DOI: 10.1016/S0019-9958(65)90241-X

Zadeh, L. A. (1993). Fuzzy logic and approximate reasoning. *Synthese, 98*(1–2), 149–155. DOI: 10.1007/BF01064189

Zadeh, L. A. (1994). Fuzzy logic, neural networks, and soft computing. *Communications of the ACM, 37*(3), 77–84. DOI: 10.1145/175247.175255

Zhang, W.. (2022). Advancing surgical training with AI-powered simulations. *Surgical Education Quarterly, 29*(2), 87–102.

Zhang, Y.. (2021). Black-box explanations in AI: Towards human-centric AI systems. *Journal of Artificial Intelligence Research, 70*, 701–743.

Zheng, A., & Casari, A. (2018). *Feature engineering for machine learning: principles and techniques for data scientists.* O'Reilly Media, Inc.

Zheng, Y., & Wang, Y. (2023). High-resolution reflection seismic imaging to reveal subsurface geologic structures of a deep geothermal reservoir. *Geophysics, 88*(5), WB37–WB43. DOI: 10.1190/geo2022-0494.1

Zhou, S. K., Greenspan, H., Davatzikos, C., Duncan, J. S., Van Ginneken, B., Madabhushi, A., Prince, J. L., Rueckert, D., & Summers, R. M. (2021). A review of deep learning in medical imaging: Imaging traits, technology trends, case studies with progress highlights, and future promises. *Proceedings of the IEEE, 109*(5), 820–838. DOI: 10.1109/JPROC.2021.3054390 PMID: 37786449

Ziegler, S., & Burfeindt, M. (2025). Frequency beamforming-enhanced DBIM for limited-aperture quantitative imaging. *IEEE Transactions on Antennas and Propagation, 73*(2), 1173–1184. DOI: 10.1109/TAP.2025.3526904

Zimmermann, H. J. (2001). *Fuzzy set theory and its applications*. Kluwer Academic Publishers. DOI: 10.1007/978-94-010-0646-0

About The Contributors

Mohamed Hammad, Received my Ph.D. degree in 2019, at the School of Computer Science and Technology, Harbin Institute of Technology, Harbin, China. I'm an assistant professor in the Faculty of Computers and Information, Menoufia University, Egypt. I'm currently a researcher in the EIAS Data Science Lab, College of Computer and Information Sciences, Prince Sultan University. I'm an ACM Professional Member. My research interests include Biomedical Imaging, Bioinformatics, Cyber Security, IoT, Computer Vision, Machine Learning, Deep Learning, Pattern Recognition, and Biometrics. I have published more than 50 papers in international SCI-IF journals. Furthermore, I have served as an Editor Board member in PLOS ONE journal, an Editor Board member in BMC Bioinformatics journal, an Associate Editor in IJISP, a Topics Board editor in Forensic Sciences (MPDI) journal, a guest editor in many international journals such as IJDCF, an editorial board of World Journal of Methodology, Sensors (MDPI) and Information (MDPI). Reviewer of more than 500 papers for many prestigious journals and listed in the top 2% of scientists worldwide (According to the recently released list by Stanford University USA in 2022, 2023 and 2024).

Hatem Abdulkader received the B.Sc. and M.Sc. (by Research) degrees in electrical engineering from the Faculty of Engineering, Alexandria University, Egypt, in 1990 and 1995, respectively, and the Ph.D. degree in electrical engineering from the Faculty of Engineering, Alexandria University, in 2001, specializing in neural networks and applications. He is currently a Professor with the Information Systems Department, Faculty of Computers and Information, Menoufia University, since 2004. He has worked on a number of organizations. He has contributed more than 150 technical articles in the areas of neural networks, big data, database applications, information security, and internet applications.

***

Sadique Ahmad is a research scholar with expertise in Deep Cognitive Modeling for Multiple Goals, such as big data analysis, intelligent decision-making, and object detection. His work has taken him to national and international institutes, providing him with a global perspective and adaptability. He has more than ten years of teaching and research experience in national and international institutions, including China, Pakistan, and Saudi Arabia. Dr. Ahmad has published over 88 peer-reviewed research articles and review papers in Web of Science (SSCI, ESCI) and Scopus (Q1, Q2, Q3) journals. He has received 946 citations and an h-index of 18, while the i10 index is 27. His is also a guest editor of special issues in the journals "Frontiers in Computational Neuroscience" and "Frontiers in Oncology". He is also the editor of four books on deep cognitive modelling in education, object detection, healthcare, and medical imaging.

Pawan Kumar Badhan is an Assistant Professor in the IT Department at Pyramid College of Business and Technology, Punjab, India. He has over 14 years of experience in teaching, research, and industry-oriented training. Pawan holds a Master of Computer Applications from Punjabi University, Patiala, and has a strong academic foundation in computer networks, machine learning, deep learning, and software engineering. His research focuses on leveraging machine learning and quantum computing for applications in IoT networks, anomaly detection, and cybersecurity. Pawan has published multiple international research papers, book chapters, and patents, including works on early disease detection and IoT-based innovations. He has served as a reviewer for leading journals such as Neural Computing and Applications (Springer) and Biomedical Signal Processing and Control (Elsevier). His contributions to academia include organizing workshops, industrial visits, and seminars to bridge the gap between theoretical knowledge and practical applications. Pawan is also a recipient of several accolades, including being recognized as an "Out-of-the-Box Thinker" by Harper Collins Publishers, India.

Shanmugasundaram Hariharan received his B.E degree specialized in Computer Science and Engineering from Madurai Kammaraj University, Madurai, India in 2002, M.E degree specialized in the field of Computer Science and Engineering from Anna University, Chennai, India in 2004. He holds his Ph.D degree in the area of Information Retrieval from Anna University, Chennai, India. He is a member of IAENG, IACSIT, ISTE, CSTA and has 17 years of experience in teaching. Currently he is working as Professor in Department of Computer Science and Engineering, Vardhaman College of Engineering, Hyderabad, India. His research interests include Information Retrieval, Data mining, Opinion Mining, Web mining. He has to his credit several papers in referred journals and conferences. He also serves as editorial board member and as program committee member for several international journals and conferences.

Gaganpreet Kaur is currently working as an Professor at Department of Computer Science & Engineering, Chitkara University, Punjab, India. She has obtained Ph.D (CSE),from I.K. Gujral Punjab Technical University, Jalandhar, M.Tech (CSE), from I.K. Gujral Punjab Technical University, Jalandhar and B.Tech (CSE), from Kurukshetra University. She has rich teaching experience of 18+ years. She has published more than 100+ research articles in various peer reviewed, SCI, Scopus Journals, International and National Conferences in the field of computer science. She has published more than 17 Indian patents. She has authored a book on biometrics. She has guided a Ph.D scholar and 38 students have M. Tech students. She has acted as a guest editor for special issue in various Special Issues.

A.V.Senthil Kumar is working as a Professor and Director, PG and Research Department of Computer Applications, Hindusthan College of Arts & Science, Coimbatore, Tamilnadu. To his credit he has industrial experience for five years and teaching experience of 27 years. He has also received his Doctor of Science (D.Sc in Computer Science). He has to his credit 50 Book Chapters, 225 papers in International and National Journals 64 papers in International Conferences in International and National Conferences, and edited 14 books and 2 Text books. He is as Associate Editor of IEEE Access. He is an Editor-in-Chief for many journals and Key Member for India, Machine Intelligence Research Lab (MIR Labs). He is an Editorial Board Member and Reviewer for various International Journals. He is also a Committee member for various International Conferences.

Rohaya Latip, She is currently an Associate Professor at the Department of Communication Technology and Network, Faculty of Computer Science and Information Technology, Universiti Putra Malaysia. She holds a Ph. D in Distributed Database and Msc. in Distributed System from Universiti Putra Malaysia. She graduated her Bachelor of Computer Science from University Technology Malaysia, Malaysia. She was the head of Department of Communication Technology and Network from 2017 until 2022. She served as an Associate Professor at Najran university, Kingdom of Arab Saudi from 2012 until 2013. She was the Head of HPC section in Universiti Putra Malaysia (2011-2012) and consulted the Campus Grid project and the Wireless for hostel in Campus UPM project. Her research interests include Big Data, Cloud, Fog and Edge Computing, Network Management, and Distributed Database. She has published more than 150 papers in international and national journals, proceedings and posters.

Sajad Ahmad Assistant Professor I Public Administration I Political Science I Police Administration I Research Methodology Introduction I am a dedicated academician and researcher with a Ph.D. in Public Administration, committed to advancing knowledge through rigorous research and scholarly publications in esteemed Scopus-indexed journals. My academic journey is enriched by active participation in international conferences, contributing valuable insights to the field. My passion for education and research has driven me to create meaningful contributions in Public Administration, Political Science, Police Administration, and Research Methodology. Academic and Professional Journey My professional journey in academia began with a strong inclination toward teaching and research. Over the years, I have refined my pedagogical approaches, integrating innovative teaching methodologies with research-driven insights to enhance student learning outcomes. My academic career has been shaped by various teaching and research experiences, which have provided me with a comprehensive understanding of higher education's dynamic landscape.

Vaishali Mishra is a Senior Product Manager specializing in Data & AI, with expertise in data quality, governance, and analytics. She has extensive experience building scalable data platforms, AI-driven solutions, and cloud migrations. She leads strategic data initiatives at Expedia Group, optimizing data management and AI capabilities. Vaishali is also an industry speaker and a researcher. Her work has significant real-world applications, particularly in the travel, healthcare and biomedical industries.

Hemlata Parmar, a luminary in the field of computer Science and Engineering. Her Journey is defined by a constellation of achievements from prolific publications and patents to enlightening guest lectures. Yet, her brilliance is matched only by her unwavering commitment to human ethics. With a heart as vast as her intellect, she champions women's empowerment, bridging the gap with empathy and insight. In every endeavor, she embodies the perfect fusion of academic process and compassionate advocacy, inspiring generations to come.

Omar S. Saleh earned his Bachelor of Science (B.Sc.) in Computer Engineering from the University of Technology-Baghdad in 2006. He later obtained his Master of Science (M.Sc.) in Software Engineering from Staffordshire University-APU, Malaysia, in 2012, and completed his Ph.D. in Information Technology from Universiti Utara Malaysia in 2024. Currently, Dr. Saleh serves as an Associate Professor at the Studies, Planning, and Follow-Up Directorate of the Ministry of Higher Education and Scientific Research in Iraq. With a strong academic and research background, he has authored and co-authored over 30 publications, including journal articles, conference papers, and book chapters. His research focuses on cutting-edge technologies such as Blockchain, Cloud Computing, Software Architecture, and the Internet of Things (IoT). In addition to his contributions to academia, Dr. Saleh is an active reviewer for several indexed journals and has played a key role in organizing prestigious international conferences, including those sponsored by IEEE.

V. Santhi has received her Ph.D. in Computer Science and Engineering from VIT University, Vellore, India. She has pursued her M.Tech. in Computer Science and Engineering from Pondicherry University, Puducherry. She has received her B.E. in Computer Science and Engineering from Bharathidasan University, Trichy, India. Currently she is working as Associate Professor in the School of Computing Science and Engineering, VIT University, Vellore, India. She has authored many national and international journal papers and one book. She is currently in the process of editing two books. Also, she has published many chapters in different books published by International publishers. She is senior member of IEEE and she is holding membership in many professional bodies like CSI, ISTE, IACSIT, IEEE and IAENG. Her areas of research include Image Processing, Digital Signal Processing, Digital Watermarking, Data Compression and Computational Intelligence.

Vijender Kumar Solanki, Ph.D. is an Associate Professor in Computer Science & Engineering, CMR Institute of Technology (Autonomous), Hyderabad, TS, India. He has more than 10 years of academic experience in network security, IoT, Big Data, Smart City and IT. Prior to his current role, he was associated with Apeejay Institute of Technology, Greater Noida, UP, KSRCE (Autonomous) Institution, Tamilnadu, India & Institute of Technology & Science, Ghaziabad, UP, India. He has authored or co-authored more than 25 research articles that are published in journals, books and conference proceedings. He has edited or co-edited 4 books in the area of Information Technology.. He received Ph.D in Computer Science and Engineering from Anna University, Chennai, India in 2017 He is Editor in International Journal of Machine Learning and Networked Collaborative Engineering (IJMLNCE) ISSN 2581-3242, Associate Editor in International Journal of Information Retrieval Research (IJIRR), IGI-GLOBAL, USA, ISSN: 2155-6377 | E-ISSN: 2155-6385 . He is guest editor with IGI-Global, USA, InderScience & Many more publishers. He can be contacted at spesinfo@yahoo.com .

Kyla is a colleague of Space Foundation and NASA employees and a member of a military family with ties to the armed services (such as the Marines, Navy, and Army...) is a United Nations Global Ambassador for Peace (for the African Union 6th Region, African Diaspora) for women's empowerment, economic development, financial investment, and sustainable development goals achievement. Including an industry practitioner and DM (Doctor) in Management and Organizational Leadership. She focuses on Change Management and mixed methods empirical research as a Management Consultant, for continuous improvement, business transformation, and recovery during and after internal/external crisis to nations, institutions, and individual's lives. Her career began in December 2002 in banking and financial services at Wells Fargo Bank, N.A. (North America), Wells Fargo Financial, LLC, US Bank, and SunTrust Banks, Inc. with affiliate relationships at Wells Fargo Advisors, Wells Fargo Investments, Wells Fargo Business Banking (Commercial), Wells Fargo Student Loans, Wells Fargo Home Mortgage, and Wells Fargo Card Services. With roles in/as banker, underwriting, financial crime compliance (OFAC, MTR, CTR, SEC), loan administration, for example, and leadership roles in Credit Management & Analysis, Credit Risk, Risk Management, Cash Management, Front Office (in the Branch) and Back Office (in the operations center), Bank Products and Services (cross-selling), Lending, Suppliers/Vendor Management and Procurement, Operations, and Corporate Banking (like for hospitals and clinics). She has media and press appearance in Forbes, ABC, Fox News, Yahoo! Finance, U.S. News & World Reports, Financial Times, CNBC Arabia on Middle East and US Economic Outlook. She is an on-call Grant Reviewer for organizations like the United States Department of Agriculture (e.g. responsible for awarding funding for programs up to $65M in funding) and other institutions. Inclusive, a part-time PhD Faculty of Corporate Governance at Woolf University and WDC Leadership School UK/USA (see here http://wdc.woolf.university/faculty/93d5d684-4a7a-4f5b-afd3-44645e9ecc90), and Fellow-in-Residence and Lead Researcher at University of Phoenix College of Doctoral Studies, Center for Leadership Studies and Organizational Research (CLSOR). Also, she is a leader in various roles at investment forums, conferences, and summits where she works with Presidents of Nations, Heads of Banks, Prime Ministers, High Commissioners, and Chambers of Commerce and other business leaders to strategize and make business decisions to help businesses start-up, advance, and/or scale to improve economic development, promote social justice, and create jobs in various nations. Including at the World Business Angels Investment Forum (WBAF). An affiliated partner of the G20 Global Partnership for Financial Inclusion (GPFI), where the vision is to promote social justice, create jobs, financial inclusion, and innovative financial instruments for innovators, startups, and SMEs, helping them to scale. Dr. Tennin is also affiliated with the International Monetary Fund, World Bank Group, and Parliamentary Network to enhance economies and various forms of businesses, and has spoken on stages with Ms. Rula Ghani, First Lady of Afghanistan, Princess Dina Mired of Jordan, Ms. Intissar El-Sisi, spouse of Egyptian President, Dr. Mariem Mint Dah, First Lady Of Mauritania, Dr. Madeeha bint Ahmed al Shibaniyah, Minister of Education (Oman), and Ms. Amina J. Mohammed, Deputy Secretary-General of the UN & Chair of the UN Sustainable Development Group. Dr. Kyla is an OB/GYN MD track turned DM (doctor) with background in biology, biotechnology, and organic and biochemistry, with major emphasis in women and gender studies, has project management and lean six sigma certifications, a Master of Business Administration (MBA), and is a Ph.D. Fellow at the Global Peace Institute, headquartered in the United Kingdom. She contributed to the accreditation process at the University of Phoenix College of Doctoral Studies and completed additional studies (to launch and scale corporations) at Harvard, Harvard Law (e.g. contracts and negotiation), Stanford, Yale, University of Cape Town, Cornell University - The Bank of America Institute for Women's Entrepreneurship at Cornell (certificate in Women's Entrepreneurship – sponsored by Bank of America), and University of Glasgow to name a few. She has personally written approximately 18 books, 3 book chapters with colleagues, 3 patents, several other intellectual properties, and nearly 85 research reports, later conceptualized into products and services for global businesses. Along with additional major book deals with approximately 8 books in the pipeline with forthcoming publishing. In areas of Financial Crimes, Fraud, Banking, Artificial Intelligence, Case Studies, Sustainable Development (ESG), and forthcoming of Teaching Notes and Textbooks (see a few here). Pipeline book deals and current book deals with IGI Global Publishing, NOVA Publishers, Emerald Publishing, Springer, Springer Nature, Palgrave MacMillian, Wiley, Routledge, and Taylor & Francis Group (CRC Press). Dr. Tennin has presented research and posters in practitioner and academic conferences, workshops, and international webinars. She has professional experience in the financial services and investments, hospitality, restaurant, consumer products/beauty, and healthcare industries, with 95% of her experience spent in financial services and higher education. Dr. Tennin has 23 years of experience in finance, financial services, and banking industries and 17 years owning her own global conglomerate corporation, which she built while homeless, to recover for self-sufficiency, to recover financially, and create jobs for people. Operating in 31 countries, with 24 locations, on 6 continents, and 2 world headquarters. She was awarded over 30 corporate awards, primarily in the financial services industry, mentored entrepreneurs for 22 years with global organizations and world-renowned corporations and Development Banks and Commercial and Retail Banks as sponsors, and currently owns a business academy, women's entrepreneurship academy, and women's entrepreneurship bootcamp program where entrepreneurship, resilience, finance, pricing and tax strategy, research methodology and data analysis for consultants, personal and business credit, corporate governance and risk management, digital transformation, property ownership (real estate (e.g. commercial and residential)), and other courses are taught. Development Banks like Jamaica Development Bank, Qatar Development Bank, Asian Development Bank

(ADB), and European Bank for Reconstruction Development Bank for corporations and SMEs. Her areas of expertise and research interests are financial and revenue strategy, entrepreneurship (SME), cosmeceuticals, economic development, emerging markets, multinationals, corporate governance, well-being/burnout, and women's empowerment.

Duc-Tan Tran is an Associate professor and Vice Dean of Faculty of Electrical and Electronic Engineering, Phenikaa University. He has published over 150 research papers. His publications received the "Best Paper Award" at the 9th International Conference on Multimedia and Ubiquitous Engineering (MUE-15), and International Conference on Green and Human Information Technology (ICGHIT-2015). He was the recipient of the award for the excellent young researcher from Vietnam National University in 2008, Hanoi and the third prize in the contest "Vietnamese Talents" in 2008. His main research interests include the representation, processing, analysis, and communication of information embedded in signals and datasets. He serves as a TP Co-chair, technical committee program member, track chair, session chair and reviewer of many international conferences and journals.

Index

Symbols

A

B

C

D

F

H

I

M

P

R

S

www.ingramcontent.com/pod-product-compliance
Lightning Source LLC
LaVergne TN
LVHW080856300926
849162LV00003B/439